Advanced Hypnotherapy Scripts

EXPANDED EDITION

A Collection of Over 100 Hypnosis Scripts From Inductions & Deepeners to Treatments, Meditations; Therapeutic Processes & Trance Termination & So Much More...

Includes expanded inductions and 'quit smoking scripts and strategies' previously released in 'Advanced Hypnotherapy Scripts Collection: Quit Smoking Scripts'

With

Ericksonian Hypnotherapist

Dan Jones

GHR (Reg.), GQHP, DHypPsych (UK), Dip.I.Hyp.E.Psy.NLP Prac (BHR)

D.NLP, HypPrac, BSYA (B.D, Cur.Hyp, H.Md, Zen.Md) MASC (Relax, NLP)

Contact the author:

www.DanJonesHypnosis.com

Second Edition 2014

Published By Neuro Publishing

Copyright © Daniel Jones 2011

Daniel Jones asserts the moral right to be identified as the author of this work

All rights reserved. No part of this publication may be reproduced, stored in a retrieval system, or transmitted, in any form or by any means, electronic, mechanical, photocopying, recording, or otherwise, without the prior written permission of the publishers or author.

ISBN: 978-1505809992

2SECOND EDITION2

Contents

Introduction	9
HOW TO INDUCE A TRANCE	13
AN EXAMPLE OF AN OVERT CONVERSATIONAL INDUCTION	15
AN EXAMPLE OF A PATTERN INTERRUPT INDUCTION	15
AN EXAMPLE OF A METAPHORICAL INDUCTION	16
AN EXAMPLE OF A DIRECTIVE INDUCTION	17
AN EXAMPLE OF A CONFUSION INDUCTION (USED WITHIN A STORY)	18
NATURALISTIC INDUCTIONS	18
OTHER USEFUL WAYS FOR BEGINNERS TO INDUCE TRANCE AND DO EFFECTIVE THERAPY	20
AN INTRODUCTION TO ERICKSONIAN LANGUAGE PATTERNS	21
Agreement set	21
Linking Suggestions	22
Embedded commands and suggestions	23
Double binds	24
Metaphors	25
Presuppositions	26
STRUCTURING A THERAPY SESSION	26
Understanding the Scripts in This Book	29
Hypnotic Inductions, Deepeners & Priming	39

Induction Introductions — 39
- *Induction introduction for clients that have been hypnotised before* — 39
- *Induction introduction for clients that haven't been hypnotised before* — 40

Targeting specific types of client — 41
- *Logical analyser* — 41
- *Daydreamer* — 41
- *Motivated & willing to follow instructions* — 42
- *Passive – expects the work to be done to them* — 42
- *Polarity responder* — 42

Inductions — 43
- *Journey along a beach* — 43
- *Journey through a forest* — 44
- *Journey through an art gallery* — 45
- *Journey through a country meadow* — 46
- *Confusion – lifting and moving the arm* — 47
- *Confusion induction* — 48
- *Early learning set induction* — 49
- *Eyes open eyes closed induction* — 50
- *Elman induction* — 52
- *Four seasons induction* — 53
- *Staircase induction* — 54
- *Conscious/unconscious separation induction* — 55
- *Body scan induction* — 56
- *Arm catalepsy induction* — 57
- *Three things induction* — 58
- *Self-suggestion induction (good for people that over analyse* — 60

Deepeners — 61
- *Staircase* — 61
- *Pictures in pictures* — 61
- *Raising and lowering the arm* — 62
- *Down in a lift* — 62

    Gazing into space   *63*
    Using silence   *64*
  PRIMING   64
    Priming to overcome psychological difficulties   *64*
    Priming to overcome physical difficulties   *66*

Therapeutic Scripts, Processes & Techniques   67

  SMOKING HABIT THAT CURRENTLY SERVES A PURPOSE (LIKE FOR MANAGING STRESS OR BOREDOM)   67
  SMOKING HABIT THAT NO LONGER SERVES A PURPOSE   68
  SCRAMBLING   69
  STOP SMOKING SCRIPT FOR SMOKERS THAT HAVE STRUGGLED TO QUIT BEFORE   72
  QUIT SMOKING REPROGRAMMING TECHNIQUE   81
  DISSOCIATION AND REINTEGRATION QUIT SMOKING PROCESS   83
  WEIGHT LOSS – TRANCE-LIKE EATING, CAN'T STOP EATING   86
  WEIGHT LOSS – METAPHORS FOR PUTTING IN HARD WORK, GETTING A REWARD, FINDING THE INNER BEAUTY, ACHIEVING YOUR PREFERRED PHYSICAL IMAGE   87
  WEIGHT LOSS – COMFORT EATING OR EMOTIONAL EATING   87
  WEIGHT LOSS – LIMITED TASTES; EATING TOO MUCH 'JUNK FOOD'   88
  WEIGHT LOSS – WANTING TO EXERCISE MORE   89
  DEPRESSION CAUSED BY ISSUES IN THE PAST   89
  DEPRESSION CAUSED BY ISSUES IN THE PRESENT   90
  DEPRESSION CAUSED BY WORRYING ABOUT THE FUTURE   91
  DEPRESSION WHERE THE CLIENT CAN'T WORK OUT WHY THEY FEEL DEPRESSED   92
  WORRYING ABOUT EVENTS   93
  CERTAIN SITUATIONS TRIGGER PANIC   94
  MANAGING PAIN OF AN INJURY THAT STILL NEEDS LOOKING AFTER   94
  MANAGING CHRONIC PAIN – LIKE ARTHRITIS   95
  REMOVING PHANTOM LIMB PAIN   97
  PERFORMANCE ENHANCEMENT – GETTING IN THE ZONE   98

Performance enhancement – improving ability	99
Learning to relax	101
Removing a phobia with no known origin	102
Removing a phobia or PTSD with a known origin	104
Obsessive Compulsive Disorder Relief	106
Boosting confidence	108
Enhancing motivation	110
Improving self image	111
Setting and achieving goals	113
Metaphorical 'life changer'	114
Overcoming insomnia	115
Overcoming impotence	116
Overcoming enuresis (bed wetting)	118
Life reprocessing technique	119
Comfortable birth	120
Overcoming fibromyalgia	120
Acne removal	121
Overcoming anger	122
Improving the nervous system	123
Improving the endocrine system	124
Improving the digestive system	125
Improving the circulatory system	126
Overcoming Irritable Bowel Syndrome (IBS)	126
Boosting the immune system & fighting infections	127
Increase fertility	128
Treating psoriasis	129
Self awareness process	130
Self acceptance process	131
Enhance intuition	132
Forgiveness process	133
Finding inner peace process	134

DEVELOPING ASSERTIVENESS	135
OVERCOMING ADDICTION	136
HAY FEVER RELIEF	137
ENHANCE ORGASMS	138
OVERCOME PREMATURE EJACULATION	139
OVERCOMING INSECURITY	139
HEAL ULCERS	140
ENJOY EVERYDAY SENSORY EXPERIENCE	141
OVERCOMING PROCRASTINATION	142
ENHANCING CONCENTRATION	143
FOCUS MEDITATION	144
LOVING KINDNESS MEDITATION	145
OVERCOMING NAIL BITING	146
CHANGE THAT PROBLEM NOW (GENERIC PROBLEM RESOLUTION SCRIPT)	147
WARTS REMOVAL	147
HELP WITH MEETING BASIC EMOTIONAL NEEDS AND USING THEIR INNATE SKILLS CORRECTLY	149
To give and receive attention	*149*
Mind body connection	*149*
Purpose and goals	*150*
Belonging to the wider community	*151*
Stimulation and creativity	*151*
Understood and emotionally connected to others	*152*
A sense of control and independence	*153*
Tolerating uncertainty	*154*
Helping the client challenge their emotional thinking	*154*
Transferring natural resources, skills and strengths	*155*
Managing attention	*155*
Using creative thinking and thinking of alternative viewpoints	*155*
Reorienting Back to the Room	**157**
EXITING TRANCE NOW	157

ALLOW THE CLIENT TO EXIT TRANCE WHEN THEY ARE READY	158
ALLOW THE CLIENT TO DRIFT OFF TO SLEEP	158
Treatment Ideas	**161**
GATHERING INFORMATION	161
TASK SETTING	162
GETTING EMOTIONAL NEEDS MET	164

Introduction

Since getting involved in the field of hypnosis over 20 years ago I have been fortunate enough to have helped people with a wide range of problems from clients wanting to quit smoking to people with anxiety and depression and pretty much everything in between.

In the early days I used to get a client's history and information about the problem and what they would like as the solution then by the time they came for the first session I would have written a script and developed treatment ideas. Before the second session I would have contact with the client by phone to see how they are getting on and what changes they have noticed, what improvements and what they still want to improve. From this information by the second session I would have written a second script for the client to help them to continue moving forwards.

Writing all of these scripts became quite time consuming so I began to purchase books of scripts. None of these scripts matched my clients exactly but they gave me ideas which I could use to mix and match for individuals I was working with.

I stopped using scripts many years ago but still think they have their place as all Hypnotherapists are unique and some like scripts whereas others aren't so keen on the use of scripts and many people like to read scripts for inspiration and ideas and to learn.

All the scripts in this book are based on my interpretation of Ericksonian Psychotherapy and Hypnotherapy. Ericksonian Therapy isn't about always being covert and indirect or all about using stories and metaphors and complex hypnotic language. It's about getting the balance right. Erickson frequently used direct suggestion and direct inductions. The Ericksonian Approach is about taking what works with each client. If the client needs to be worked with in a direct manner then do this, if they need to be worked with indirectly use this instead.

I like to work somewhere in the middle. From over 20 years' experience I have found that starting with indirect techniques to prime ideas, concepts and mental processes and begin to activate these without the client realising that that is what I am doing so they don't resist it and then moving on to being more direct with indirect techniques being used to compliment the direct work is what works best for me; and this is what the scripts in this book are like. They will involve indirect techniques and direct techniques to enhance effectiveness.

Since teaching hypnosis to others I have been repeatedly asked if I can create scripts for other therapists that they can use with their clients. Everything I'm sharing with you in this book has been used effectively with my own clients and with clients of others.

I was initially reluctant to write a book of 'scripts' but as time went on I realised that many training Hypnotherapists found the use of scripts; especially scripts they knew worked gave them confidence and this confidence influenced their effectiveness with clients.

One of the reasons for being reluctant to write a book of scripts is that it may encourage people to stop paying attention to clients and to stick rigidly to what is written in the script as if it is 'the truth' and the way you have to do things.

From this concern I have decided to include a little more information rather than just the scripts to encourage people; especially those new to the field of Hypnotherapy that are perhaps going through training to look at the client as a whole and integrate the scripts into how you work with clients rather than have the script as the starting point and trying to make clients fit the scripts.

I hope you enjoy this book and can use the scripts, strategies and ideas in your work, creating positive change in people's lives. If you find this book helpful, it would be highly appreciated and valued if you could write a review, or create your own self-hypnosis YouTube videos from the scripts in this book and share them (if you let me know about your self-hypnosis videos on my Facebook page (www.facebook.com/DanJonesHypnosis) I will share them with others).

Before embarking on using any of these scripts in your own private practice there are some useful pointers to keep in mind:

1. **Some clients come to therapists because they have been told to; not because they have chosen to.** With these clients they are

more likely not to carry out any tasks set and may come in with such a strong mind-set that they don't need your help or don't want to change or that they want the therapy to fail so that they can say they 'tried' hypnosis and it didn't work. With these clients it is difficult to help them to change using a script because they have no intention of changing. A large number of these people can be helped with Hypnotherapists that are not using scripts but instead are responding moment by moment to the client as an individual. It doesn't mean that in the future these clients couldn't come back to you to be helped when they are ready it's just now isn't the right time for them.

2. **You don't 'fix' clients they 'fix' themselves.** You are a guide for your client. If they come to you wanting help you can give them ideas and information but for them to change they need to do the work. For example if you set a task and they don't carry it out that is their choice. For lasting change the client needs to feel empowered and in control of their own situation not reliant on a therapist so they need to do the work.

3. **Change isn't always instant.** For lasting change sometimes other things need to happen outside the therapy room. For example someone may want to quit smoking but may smoke when they get stressed. They may expect 'you' to stop them smoking instantly but the thing causing them stress may not yet be dealt with or they may not have mastered relaxing in a new way. This can all take time and so if they just quit at the end of the session they are likely to fail because once they get stressed they will resort to what they know and start smoking again. If on the other hand they practice a relaxation technique and also deal with the situation that is causing them stress then they will be in a better place to quit. So in many cases the client may not get the result until months later when the time is right. I worked with someone that had a fear of success and had a dream about what she wanted to do. It was going to take some hard work on her part to make the dream a reality and her fear of success came from a lack of confidence which was due to a lack of knowledge and skill about what she needed to know. After the session she didn't think things had changed; a few weeks later she still maintained 'it didn't work' because she was still nervous about the success she wanted. She then 'forgot' about our session and just

carried on with her life; following her dream; learning more and becoming more skilled. Then a year later a friend of hers contacted me and told me she is living her dream. It all played out as I had expected. She went through all the stages I had expected and built on all the seeds sown in the single Hypnotherapy session that we had had. I had explained to her friend just after the session what I had done (they were a training Hypnotherapist and observed the session for training purposes and we discussed the session afterwards) and what I expected to happen. The one thing I couldn't answer was how long it would take her to go through all the necessary stages as she would have to find opportunities to attend courses and learn new things.

4. **The scripts in this book are best used as guides.** Every client is an individual so the best way to use and read the scripts is to keep the essence of the script whilst adapting the wording to suit each client. You can also mix and match scripts to include what would be best suited to the client you are working with. So if a specific suggestion or metaphor or induction or task seems like it would be good for a client you can use it and edit it into the 'main' script you are using.

5. **Make sure your client has an understanding of what hypnosis is and how it works.** Make sure they understand Hypnotherapy isn't a 'quick fix' therapy. I often ask clients how quick they expect the changes to take before they begin to notice the improvements. This gives me the opportunity to discuss any unrealistic expectations the client may have. So if I know a client is unlikely to be a non-smoker instantly (which in lots of cases they can be) because they have other issues to sort out and they tell me they expect to instantly be a non-smoker then I would discuss this with the client until we get some agreement on how long it is likely to realistically take, what stages the client may well go through and if they may have a relapse I will discuss with them how they will be able to have a few relapses, what they can learn from them and how they can get back on track again to move on. I talk about how in the session we lay down new patterns of responding in the brain and that these new patterns need strengthening and reinforcing to make them become habitual and some of the patterns may need specific events to happen before they become active. I often mention that sometimes suggestions and

ideas need time to seep through from the unconscious to the conscious mind because they need time to be understood and incorporated fully before they can be used.

How to induce a trance

Hypnosis can be induced by focusing attention (it could be on a spot on the wall, or on a thought, or on a rhythm, or on almost anything else).

Hypnosis or trance states can be induced in many different ways:
- By confusion followed by a solid suggestion.
- Pattern interruption, (like handshake inductions) these fire the reorientation response as the correct pattern isn't happening so they take their cue on what to do next from the hypnotist.
- Shock inductions (like most stage hypnotists do, these set off the reorientation response).
- Relaxing the muscles (which are part of the process for falling asleep).
- Deepening rhythmic breathing (part of the process for sleep).
- Visualisation (part of falling asleep and dreaming).

Everyone uses hypnosis all of the time. When people think about winning the lottery and what they would do with the money, they are visualising which induces a light trance. Smokers may go into a trance when they focus on the cigarette they are having and they take deep breaths as they allow their muscles to relax. When people have cravings they enter a trance as they become so intensely focused on what they crave. When people get angry they focus on what is causing the anger. When people get depressed they focus on worrying and negative thoughts.

Hypnosis CAN be induced in anyone

It used to be thought that not everyone could be hypnotised but this was because in the past hypnotists would use a script that was the same for each person. This is something that is often done when researching hypnosis because it means that everyone gets the same treatment so that the treatment can be evaluated.

This didn't work on everyone because people are all different. For example, some people might feel uncomfortable with an induction that guides them down in an elevator so they won't respond by going into a trance. Now well-trained Hypnotherapists will tailor the induction to the specific client and let clients go into trance in their own way.

To do hypnotic induction's you need to either recreate stages leading to dreaming sleep or recreate the state of not knowing what is happening next causing the reorientation response (the response created when a car backfires or you hear your name called or you get lost and suddenly pay lots of attention to finding anything familiar to latch onto to know you know where you are again).

Recreating stages of sleep could be a relaxation induction getting the client to relax their body perhaps starting with their head and working through their body; then relaxing their mind by getting them to think of something pleasant. Or it could be getting them to imagine something relaxing. Or getting more of their attention focused inwardly in some other way.

Recreating a state of not knowing what is happening next could be done by interrupting a pattern of behaviour, or causing confusion.

Some types of induction are:

- Conversational (overt & covert)
- Pattern interrupt
- Embedded-meaning/metaphorical
- Confusion
- Directive

Conversational inductions are induction's that initially start with an ordinary conversation. They involve embedding suggestions and utilising on-going experiences or events to induce a trance. It could be embedding suggestions in a conversation or feeding back what a client says to deepen their experience.

An example of an overt conversational induction

As you *sit back and begin to feel comfortably relaxed (Embedded command)*, I would like you to *let those eyes gently close*…that's right…recognising that with those eyes closed you can *go inside very pleasantly*, accessing memories, past experiences or other meaningful events, times gone by when you felt good…now I'd like you to *take two deep, refreshing breaths* and as you release that second breath you can *drift even more deeply* into a satisfying a pleasant state of relaxation…etc.

An example of a pattern interrupt induction

(Interrupting the pattern of a handshake)

Hi, I'm Dan (hand goes out; clients hand comes to meet it. I take it with my opposite hand, raise it with palm facing clients face then slowly start it moving to their face)…and as that hand continues to move closer to your face all by itself you can begin to notice the change in your vision…and as the vision changes you can notice how heavy those eyelids are getting…and you won't go all the way into a trance until that hand comfortably touches the face…etc.

An example of a metaphorical induction

(A metaphorical or embedded-meaning induction would be to tell a story and use embedded commands and metaphors for going in to trance)

An example used in a staff meeting to get the staff working together again:

One-day snow white decided that she wanted to go on a walk, she didn't often go out far from her home as she was unsure what she would find in the deep, dark forest. Snow white left on a path right outside her front door. The path was covered by trees arching high over head; either side of her was deep, dark forest. Snow white stuck to the path walking through the shimmering beams of light that flickered down through the trees above. As she continued to...*follow this path*...she was aware of the rhythmic beat of her feet on the ground and the sounds of birds in the trees and the rustling of leaves as the wind blew a breeze. She continued to wander and at times found her mind wonder about why she set out on this journey...after walking for a while she found herself smile as she saw a house in the distance. The house was in a clearing in the forest that was bright and cheerful. There were plants of many varieties and many flowers surrounding the house. As snow white reached the clearing she could feel the calm, warmth from the sun on her skin. Snow white could hear voices coming from the house and the closer she got the more she could tell that the people inside the house were disagreeing with each other. Snow white approached and asked one of the people what was wrong. Grumpy explained that they used to all go to work singing and dancing with enjoyment but now they seem to have forgotten how to **work as a team.** Grumpy explained that they used to push together...*pull together*...axe together...*all together*...but now they found that they couldn't. When one pushed another pulled and no work got done. Snow white asked what they do and was told that they are the team that digs and lays the foundations for new buildings. She asked them why they decided to do that work. She was told that you see buildings standing and feel proud because you know that they are standing because you built the foundations well, it makes you proud of all that hard work you did...snow white decided to tell the little people a story about a centipede that kept falling over its legs. The centipede asked a friend how he manages to walk without falling over. He was told to just...*relax*...and let all the legs...***work together***...not keep thinking about which leg should do what and when. This made no sense to the dwarves so they decided to forget what snow white said and just enjoy her company. Before snow white left she asked who made such a lovely garden. The dwarves said they all worked at it and that many of the plants have survived some harsh winters. At the end of the day snow white said good bye to the dwarves. She got right up and left. As she left she was amazed by how much happier and healthier they were starting to become. Something had happened that they were learning from which looked like it made them healthier and made them work out their differences, sneezy had stopped sneezing, grumpy was happy, bashful had clear skin and no hint of red, and all of the others had noticed improvements too. This made snow white happy as she skipped away from the house up the path leaving her adventure behind like a dream that got more out of reach like a name on the tip of your tongue as she approached her home pleased with her mini adventure, then walked through her gate and, finding it was all a dream she...***opened her eyes***...

An example of a directive induction

(A directive induction is an induction where you tell the client what to do)

I'm going to shake your hand three times...the first time your eyes will get tired...let them...the second time they'll want to close...let them...the third time they'll lock and you won't be able to open them...want that to happen, and watch it happen...now...1...2...now close your eyes...now 3...and they're locked and you'll find they just don't work, no matter how hard you try...the harder you try the less they'll work...test them and you'll find they won't work at all...

An example of a confusion induction (used within a story)

One afternoon a woman set out looking for her friend's house. She was feeling tired and not really paying attention until she got halfway there and realised she'd lost her way. She decided to check her map for directions so holding her bag in her right hand she lent right down and left her drink on the ground right by her left foot using her left hand. Then reaching right across her side with her left hand to the right pocket of her bag she discovered the map wasn't there so she took her left hand out of the right pocket deciding the pocket must be wrong and that one of the pockets left must be right. She checked one of the pockets left which wasn't right and then another pocket she thought must be right leaving only one left when she discovered that wasn't right either. She now knew it must be in the pocket left which was on the right hand side of the bag. She was convinced this pocket left must be right. She checked the pocket discovering the pocket left was indeed right and took out the map. She then lent right down and with her left hand picked up the drink stood right up and left on the *rest* of her journey.

Naturalistic inductions

Probably the easiest way for a beginner to induce a trance in someone else is to use a naturalistic approach. A naturalistic approach involves talking about everyday trance states. As you talk to a client about everyday trance states they will be familiar so will rapidly start to enter trance. If you do this utilising hypnotic language the effects will be even greater.

It can be useful to write out direct scripts then change them to indirect scripts. Writing what it is that you hope to achieve and how you will achieve this. Then you can go through the script changing anything that is too direct and that might not match the client's reality to something that will.

For example, you may say '...as you approach that old wooden staircase...' which is direct and may not match the client's view of a staircase and change it to '...as you approach that staircase...' which is more general and so it allows the client the freedom to fit this into their model of reality.

To focus attention get clients talking about something that they are interested in. In the old day's hypnotists would tell the client what to think and what to focus on. To induce a trance you need to focus attention but it doesn't matter what you focus that attention on. That is one of the beauties of naturalistic induction's. Hypnotists used to use swinging watches, stroking, telling the client to look at a spot or a candle. Modern day hypnotists get clients to focus on issues, thoughts, comments, or even the process of their problem. One quick way to hypnotise a smoker is to ask them to tell you the process they go through when they smoke.

Utilise naturalistic phenomena. Anything can be used to achieve your goals. If you want to lead to a trance state you can use naturalistic phenomena leading to trance, like sleep, day-dreaming, a leisure activity. If you wanted to evoke a hypnotic phenomenon then you can use examples of times that thy have happened naturally like numbness – sleeping on an arm or holding snow, or amnesia – forgetting someone's name or being interrupted mid-sentence.

Creating responses this way will then come from within the client so they will be more powerful. It is completely different telling someone to laugh uncontrollably than reminding them of times they found themselves laughing uncontrollably, like in school in a classroom when you know you shouldn't, and the more you try to stop the laughter the more the laughter builds up, you know that feeling?

You can get the client to talk about something they enjoy doing that makes their mind wander and as they talk about it they will begin to go back into that same state of mind again. When you hypnotise someone you want to separate the conscious and unconscious mind. You can do this by confusing the conscious or marking out different messages to the conscious and unconscious mind.

Other useful ways for beginners to induce trance and do effective therapy

- **Make someone talk about their problem without using words relating to the problem, then use this to help do treatment.** This can allow you to work completely metaphorically. You can use the metaphor they give for their problem and then just get them to play out the metaphor to a positive conclusion in the clients mind. This can be useful when you don't have enough information or time to work in depth with the client.
- **Utilise everything don't think of anything as failure.** If a client doesn't give the response that you expect then utilise what they do give you and acknowledge that what they are doing is what they need to do to achieve the desired goal. For example: If a client says that they can't relax enough to go into a trance, then say 'How did you know that you needed to have a little tension there to be able to do good effective change work?'
- **Time your rhythm to rhythm of clients breathing.** This is probably one of the easiest ways to increase your effectiveness at altering someone's state. If you match their breathing and talk with the clients out breath you can begin to slow your breathing down and begin to slow down what you say and they will begin to relax deeper. This is because breathing is such a fundamental part of life that if you match it you quickly begin to build rapport with the client on an unconscious level.
- **Use fractionation.** Fractionation is a technique developed where you take the client in and out of trance repeatedly which deepens the trance each time they go inside. This can be done simply by asking the client to open their eyes then close their eyes again and go deeper. Or you could ask them to recall something; then bring them back to the room; then again have them recall something and keep repeating this. Fractionation was created because hypnotists noticed that each time clients came into a session and were hypnotised they went deeper than they had done on previous sessions. It was realised that they didn't need to have a

big gap between sessions, the same thing occurred if the client was repeatedly hypnotised during one session.
- **Feedback what the client says as suggestions.** For example:

Client: 'My left hand feels heavier than my right'
Therapist: 'Your left hand feels heavier than your right!'

By doing this you are telling the client true statements; which helps to deepen their state and you are utilising on-going behaviour and comments to lead to the desired outcome.
- **Take the client to the future to when they no longer have the problem and ask 'what did I do that helped you?'** The psychiatrist Milton H Erickson MD would often take clients to the future then ask what he did to help them. After he did this and they told him how he cured them he would bring them back to the present and do what they said he did to cure them. It is a strong belief of all the top therapists in the world that people have the resources they need to heal themselves they just need guidance and assistance in accessing that healing power.

An Introduction to Ericksonian Language Patterns

Agreement set

The 'Agreement set' is a language process that gets agreement from the client. This agreement deepens rapport; makes it harder for the client to disagree in the future and build client responsiveness.

The usefulness of this is that if you have the client in a state of mind where they are agreeing to what you are saying and are being responsive then when you say something they may not entirely have agreed with previously they are likely to do so now.

You can get agreement by having the client say or think yes or no. So for example you could be recapping information on a referral form with the client. 'So your name is... and your address is... and your date of birth is...' and that has already got three agreements when they tell you that information is correct. You could tell them 'You wouldn't expect to go into a trance until you are ready?' Which is likely to get a 'no' response but that 'no' mean 'yes I agree with you'.

Any time you are going to say something you want agreement with it is best to say two or three things first that you know they will agree with. Make sure these things are truisms. So you may say 'You have come here today...and you would like me to help you...and you've been relaxing for a little while now...(all truisms getting 'yes responses') and you can make the necessary changes in your own unique way (likely to also be agreed with in context with the other yes's)'

Linking Suggestions

Linking suggestions are suggestions that build on each other. They link things together that may not always be connected in reality; making it seem like because one thing has happened or is true the next thing should also then happen or be true.

The two main types of linking suggestions are compound suggestions and contingent suggestions. With compound suggestions often the linking is done with an 'and' or with a pause and the first part is a truism 'pacing' statement whilst the second part doesn't necessarily have to be true but it leads. To start with it is often best to use pacing and leading truisms and if possible always use things that are true. Sometimes this can just come down to wording. If you say 'You will relax' this might not be true. If you say 'You can relax' this is true. Everyone can relax.

Example compound suggestions:

'You're looking at me and you can relax'
'You can relax and think of long forgotten memories'

'You can be thinking of those long forgotten memories and wonder what you can learn from them'

As you may have noticed above you can also use compound suggestions to guide people from external reality to internal reality. So above I started with what I could see (they are looking at me) which is also means they are focusing externally. Then I mentioned they can relax which begins to focus them internally. Then thinking of long forgotten memories deepens the focus on the internal reality and less on external reality. Then finally had them wonder what they can learn from this which increases the focus on having to now find something to learn.

Whereas with contingent suggestions often the linking is done with a time based word like 'before, during, after, as, while'. Again they often start with truisms and then lead the client in a specific direction (pacing and leading). Another pattern that can be used is 'Don't...until' which works well for people that are more likely to be resistant.

Example of contingent suggestions:

'You are reading this book as you hear your internal dialogue'

'As you hear your internal dialogue you can be curious about what you will be learning'

'Before that curiosity can deepen into unconscious learning you can read some of the scripts in this book'

'Don't let full unconscious learning happen until you have finished reading'

Embedded commands and suggestions

Embedded commands and suggestions are where you are marking out a part of the communication for the client to pick up on and respond to unconsciously. This could be done with a gesture, with a head movement, with a tonal shift or a touch. As long as the marking out is done consistently the unconscious mind will pick up on it.

Some example of embedded commands could be:

'I don't know whether...*you will discover*...that...*you relax deeply*...as you listen to my voice...or whether ...*you will discover*...that...*you become more fully absorbed in your internal experience with each out breath*...'

'Someone asked the other day how...*you go into a trance*...I began explaining to them the process of how...*you go into a trance*...I explained that firstly the client will be looking at me while I talk to them and probably won't notice at first how...*the breathing begins to slow down*...and as they...*begin to relax*...I continue talking to them and they...*go into a trance easily and effortlessly*...'

'I remember travelling on a plane and discovering how high up we went...it was so difficult not to just...*close your eyes and fall asleep*...with the sounds in the cabin and the lights turned down...I couldn't help myself...*drifting off*...'

Double binds

Double binds are a class of language pattern where you are offering choices to the client but it doesn't matter which choice they follow the outcome will still be the desired one. With double binds the outcomes are unconscious so the client has little chance to sabotage it. Whereas bind as a conscious choice which can easily be sabotaged.

Bind: 'Do you want to sit in this chair or that chair to go into a trance'

Double Bind: 'I wonder whether you will sit in one of the chairs or stand up or do something different as you go into a trance'

The client can avoid being caught in the bind by standing or lying on the floor or just flatly deciding 'I'm not going into a trance'. The double bind leaves it open for almost any response to occur for the person to enter a trance. 'do something different' could mean almost any response; and 'as' being used instead of 'to' means it is going to happen the question is what will you be doing whilst the trance is developing.

Some more examples of double binds:

'Will you go deeper into a trance with the sound of my voice or the spaces between my words?'

'Which hand do you think will lift as you enter a trance or do you think neither hand will move?'

Metaphors

Metaphors, stories and examples are a useful tool in Hypnotherapy because they allow you to prime for things to take place. For example if you want an arm levitation you can prime it in advance by talking about being desperate to politely get attention in class at school or reaching up to put shopping away or hailing a taxi or stopping a bus. All of these stories seed or prime the client with the idea of arm levitation making them more likely to do it as those patterns in the mind are activated.

If you wanted to help someone forget pain you could talk to them about being in a cinema and needing the toilet as they enter but getting so engrossed in the film they forget they need the toilet. Or having fun playing as a child and not noticing any bumps and bruises or cuts until later when you are home. Again these stories then activate related processes.

Stories also allow you to lay down future useful patterns. So if the answer to the problem is to be patient you can tell a story that gives this message.

If the answer to a problem is to relax you can lay down this pattern. You can be as abstract or as lateral as you like. You could lay down a 'relax' pattern by talking about a stick floating on a river that gets wider and wider and as it does the water slows down and isn't as rough and the stick floats slower and more gently. Or you could talk about a friend that had a situation at work that really was annoying them and how they overcame this. This second option would mirror the situation with a 'recognisable' alternative. You can also talk about past successes you've had with other clients in similar situations and what they did to move on.

Presuppositions

Presuppositions are where you are making an assumption that something will be happening. This can be done very overtly or ideally more covertly. So for example if you say 'How would you like to go into a trance today?' You are implying the client will go into a trance. You are expecting it. All you are asking is how that will happen not whether it will happen or not.

Structuring a Therapy Session

The model I use or keep in mind when I am doing Hypnotherapy sessions is the RIGAAR model, developed by Joe Griffin and Ivan Tyrrell within Human Givens Therapy.

The idea of the RIGAAR model is that it gives a structure to work from. It is a structure that includes all of the areas that you will need to cover when you are working with a client, but not all of the parts have to be done in order, and often you may be doing many parts of the RIGAAR model simultaneously. The idea is that you aim to DO all of the parts to gain the best results.

RIGAAR starts with **Rapport Building**, this is essential and does have to be done right from the beginning and maintained until the end. Next comes **Information Gathering**, this again really will be done throughout because the client's will always be giving you information, whether it is about what they like or dislike or about an interest or any number of other useful bits of information. Clients will be giving you plenty of information about the structure or important information about their problem.

As you get used to noticing information and it's uses it will become an unconscious process, as will rapport building, both of which you will do automatically.

The next stage in RIGAAR is **Goal Setting**; this will be done fairly early on depending on your questioning but will become clearer and more specific as you gather further information. Then comes **Agreeing a Strategy**, this may be done with or without the client's knowledge. Really it is a stage for clarifying what they want and how they are going to get it. The next stage is **Accessing Resources**; again this will take place throughout as the client offers the different resources to you, and as you discover or notice resources. Lastly is **Rehearsal**, this stage is where you help the client to imagine the future positively with their desired goals. This may be done to some extent throughout the whole process but is definitely an important part to fully do at the end so that they leave having clearly practiced a preferred future. It is this rehearsal stage that is where you will do the 'hypnosis'. This is where you will have the client mentally rehearse changes and whilst they are in a focused, relaxed, receptive state of mind you can present 'hypnotic script' information to the client when their mind is most primed to receive it. This is the time when you may explain to the client that you will guide them into a state of mind where they can rehearse future success, recap useful information from the session and where you will present useful ideas and information to them that will help them to move forward. You may then do the induction part of the script, followed by the 'therapeutic script' or the mental rehearsal (the order you do these will depend on your preference as a therapist and what is the best order for your specific client), then the 'exiting trance script'.

R-apport: Build rapport quickly, start getting a 'yes set' etc…

I-nformation Gathering: Quickly begin to gather information (Basic needs, interests, resources, etc…)

G-oal setting: Feedback what the client says they want, establish what the desired goal is (For example: 'so you want to be able to have more energy and be healthier by stopping smoking)

A-gree Strategy: 'What we are going to do is….' Get stages (For example: Firstly I'll teach you a new way to relax so you don't have to smoke, then I'll disrupt the old smoking pattern…)

A-ccess Resources: Get usable states like a 'relaxation' state, or a 'motivation' state, etc…then anchor them

R-ehearse: Future pace, vividly build an experience in their mind of a great, pleasurable future having achieved their goal and what they did to get there.

Understanding the Scripts in This Book

Before we move on to the scripts themselves I would just like to take some time to explain about how to read the scripts to your clients; how they are written; and how to practice 'saying' the script in the most effective manner.

'...' – When reading the Hypnotherapy scripts you will notice that I have added in '...'s between certain words and phrases. These dots indicate a space to pause. If the pause is for more than a few seconds I will usually give a suggested pause duration in brackets. This doesn't have to be stuck to rigidly. The best way to know how long to pause is to watch the client. If the pause is recommended to last for more than a few seconds you can judge how long the individual you are working with might need to do the internal work required of them during that pause. If it is quite a long suggested pause like a minute or more then you can pause for much longer than this if it necessary. Generally more silence allows for deeper work and more integration of the work. I usually pause until I see physical signs of a shift in internal focus for example noticing movement or a change in breathing or skin colour. In some cases where the pause is maybe 5 or 6 seconds there is some times an extra set of '...'s.

'Italics' – Words and sentences that are written in *'italics'* should be given added emphasis. You can practice what works best for you when adding this emphasis. Some people like to speak a little softer as they say these words or sentences others like to change their tone of voice a little up or down; and some people like to speak them with more clarity or slowing down their rate of speech on these words and sentences. The words and sentences that you emphasise stick in the mind of the listener like a deeper level of communication. These become embedded commands or suggestions that often go unnoticed consciously by the client yet their unconscious mind picks up on these suggestions.

Another way of adding this emphasis is to move your head as you talk so that the location of your voice is different when emphasising words and sentences compared to when you are communicating more to the clients conscious mind. The head movements could be leaning slightly to one side as you give the suggestions, or if you are reading the script to the client you could look up when giving the suggestions.

The scripts often include information that is designed to prime the client for what is to come or to activate a specific mental process prior to requiring it. This can be done using metaphors or examples or even provoking internal wonder around an idea. By priming clients before needing a response from them you increase the likelihood of them responding to what you are doing.

The best way to practice how to read Hypnotherapy scripts effectively is to record yourself reciting the scripts and then listen to the recordings. You can then make adjustments to get it just right.

These adjustments will make it just right for you and not necessarily for the client but they will begin to give you practice of adapting how you say the scripts and what definitely doesn't work for you. For example most people like scripts to be relaxing and to make them feel calm. This involves speaking with a calm voice and speaking slowly and gently and giving the client time to respond. Silence is an important part of doing hypnosis. When you listen to tracks you make you can adjust how long your periods of silence are and see what works best for you to allow you to keep up with the track. When you are face to face with a client you can observe them closely and watch for relaxation and tension and adjust your speed accordingly. You can do the same with silences. You can remain silent until you notice subtle signs of movement. These are often cues to let you know they have now done any internal work asked of them and are ready for the next step.

As you hypnotise people and give therapeutic suggestions clients become more responsive and enter a state of heightened awareness. In this state they pick up on your thoughts and emotions (although they may not be consciously aware of this). Because of this you need to be congruent in what you say and do. If you want them to relax you need to relax and speak in a relaxing tone of voice. If you want them to exit trance you need to begin to be more full of energy, and speaking in an increasingly wide-awake lively manner. If you want them to be curious or excited you need to get into a curious or excited state first and convey this through your voice.

Because clients enter a heightened state of awareness you need to 100% believe in what you are saying and doing. If you have any doubt then this doubt will be given off in your voice and in the way you present yourself which will add doubt to the client and make the work you do less effective so approach everything you do with a sense of expectation. Expect that the client has the skills and abilities to succeed. If you believe in the client they can begin to believe in themselves.

The next chapter will include a selection of hypnotic inductions and deepeners that you can use with your clients.

Many people wonder about how deep the client has to be; whether they were in a trance or not and whether they fell asleep in the session or not (which in my experience is very rare. I used to think people were falling asleep but they always came out of trance on cue and 'got better'. I then realised they just went so deep they appeared asleep. On rare occasions people have fallen asleep but it has always been people that were desperate for sleep that perhaps had insomnia and never seemed to be able to relax enough to sleep well until they were hypnotised. For these people their unconscious minds clearly prioritised sleep above any work we were going to do together).

The depth of trance in therapeutic hypnosis isn't much of an issue. The most important thing is focus and response attentiveness. That is; you are looking for the client to be following your lead and focusing on what you are wanting them to focus on. As you talk with the client you can notice whether they are responding to you or not. You can see if they are nodding to things you say; notice whether they are focusing on you or looking around the room and fidgeting; etc. If they aren't focusing well you can be direct in getting their attention by suggesting they look at you uncross their legs and put their feet flat on the floor (or any other change of posture that you can suggest). You can reflect back what they say and ask them 'is this correct' so that they have to pay attention to what you are saying. You can talk softer and quieter so that they have to pay closer attention to hear you. All of this will increase the focus and response attentiveness of the client.

In hypnosis nothing is failure just something else to utilise. If for example a client opens their eyes and they weren't supposed to then you can suggest 'that's right' acknowledging that they have just done the right thing by opening their eyes. And then follow that with a suggestion for eye closure and going deeper. If they say they don't feel comfortable closing their eyes or they keep opening their eyes then you can suggest 'or would you prefer to look up at that spot on the wall and keep your eyes wide open as I talk to you'. This instantly utilises the fact the client doesn't want to close their eyes and instead of thinking you are failing or that they are being a difficult client you are focusing their attention and if they look at that spot on the wall and keep their eyes open they are also then being responsive to you again.

As you hypnotise people it is usually best to speak slowly and clearly in time with the clients breathing so that you talk as they breathe out and you breathe in as they breathe in. This is especially useful as you guide the client into trance. As you guide them out of trance it is useful to speak on the clients in breath so that they pick up on the different pattern. Generally I like to try to stick to speaking on the clients out breath throughout the induction and 'therapy' part of the session. Nowadays I don't use scripts, part of the reason for this is that there is very little distinction in the work I do between induction and therapy. I don't normally do formal inductions so often the client doesn't know when the hypnosis begins or when the therapy begins and I respond to the ongoing feedback and behaviour of the client so I don't know what I will be saying or doing from moment to moment as it is all dependent on the client sitting in front of me.

In the next chapter you will notice that the inductions and deepeners are broken into sections; induction introduction; type of client; induction; deepener. With your clients you can just suggest they close their eyes and then just take an induction from the induction section. Or you can make it a little more specific by including an introduction section and targeting the type of client you are working with and if you want to you can include a deepener before moving onto the 'therapeutic section' where you can select one of the scripts and finally the exiting trance stage.

By breaking down the scripts into parts like this means that if you know of specific inductions you like from other books or courses you can use those instead and just add in the therapeutic section from this book.

I have started each script section that you can choose so that they follow on from previously chosen sections. You can change the wording if you decide not to include certain sections or feel slightly different wording would work best for your specific client. For example the inductions don't mention closing the eyes because this happened in the induction introduction so if you skip the introduction then you can ask the client to close their eyes before reading the induction. Likewise all the inductions end in the same way so they can flow into the next stage. Again you can change this if you have a preferred way of doing things.

Each individual induction element may appear short, but it is just an element. On average in an hour long therapy session the 'hypnosis' part is likely to be about 20 minutes long. It could be longer or shorter depending on individual clients and circumstances but this would be an average length for the hypnosis element. This 20 minute 'inner-work' process matches the 20 minute rest phase of our Basic Rest and Activity Cycle (BRAC) which is an Ultradian Rhythm we go through taking 90-120 minutes all day every day. When you take all the relevant elements they equal a hypnotherapy script that is about 20 minutes long.

Here is an example of a script created using script elements from this book. You can visit http://youtu.be/LxF3-7FWA-M where you will find a video of the hypnotherapy script below being read out so that you can close your eyes and listen along as if you were a client, and you can get an idea for how the script elements fit together to create a whole full length hypnosis session:

(Induction Introduction) Induction introduction for clients that haven't been hypnotised before

'As you...*allow your eyes to comfortably close*...I'll explain a little about what to expect as...*you enter trance*...you already have all the knowledge you need to understand how to...*enter a trance*...you have experience of...*entering a trance*...many times each day...I'm sure you can recall times you have been wandering along and discovered...*your mind begins to wonder*...to thoughts, ideas and dreams...or the experience of waiting in line as...*you start to daydream...now*...as you sit there beginning to drift into a comfortable hypnotic trance you can hear the sound of my voice and have an awareness of different thoughts and ideas that come to mind...and while you listen to the sound of my voice I'm going to talk to you...and just like in everyday situations like when you were in school listening in class and you will have occasionally drifted off into a daydream and later realised you don't know if you remember the whole lesson...many people that...*go into a trance*...remember some things and forget others...and you can remember what is important to consciously remember whilst at the same time forget what is unimportant to recall after this session...so there is no need to try to...*go into a trance*...it can develop all by itself...and you can develop that trance in your own unique way based on your own lifetime full of experiences of entering trance states easily and effortlessly...from all those times you have discovered yourself to have been daydreaming...through to those times...*you become deeply absorbed in what you are doing*...so now...with your eyes closed you can enjoy a moments quiet while...*your unconscious*...prepares to enter a resourceful trance state...that is just right for the work we are doing here today...(pause for about 30 seconds)...that's it...now as people...*enter trance*...they often begin to notice how the breathing changes...for some people the breathing speeds up a little before it settles on a relaxed rhythm...for others it just begins to...*comfortably relax*...and while the breathing helps to deepen and maintain the trance many people become intrigued to discover how much more youthful they look as their facial muscles smooth out...this smoothing and relaxing of the muscles in the face and throughout the body and the continued feeling of comfort that people feel when you come out of trance later are some of the side effects of...*entering a hypnotic trance in this way*...so while you prepare to...*enter this trance state*...I wonder what other positive side effects you will receive...and while you continue into this state...your mind will wander at times and whatever happens my voice can go with you...'

(Induction) Journey through an art gallery

'(Pause for about 20 seconds)...that's right...now...as you listen to me and continue to breathe in that way and notice certain sensations you can continue to *become more fully absorbed* with the sound of my voice or the spaces between my words while I talk to...*your unconscious*...now...and you can discover yourself entering an art gallery...and as you enter I wonder whether you can notice the way sound changes as you enter and the way the atmosphere seems to *become calmer and more relaxed*...and as you explore the gallery different art work can give you different feelings and different meanings...you may look at a sculpture and understand that it started as just a lump of stone that the artist believed could be more and knew all it needed was to have some time and attention spent on removing what isn't necessary and reshaping what is left...and different pictures will convey different meanings...and as you continue to walk deeper into the art gallery you can wonder which painting you will feel a connection with and feel drawn to...and you may want to get a sense of a number of pictures before finally settling on the one that draws you in...and as *you drift deeper* with each out breath while exploring the various paintings my voice will quieten down in the background for a minute and you can take as long as you want to explore the gallery...and when you next hear me talk you can be prepared to step inside that picture ready to search for a comfortable and familiar chair you've never seen before...(pause for about 1 minute)...that's right...preparing now to step into that picture...and as you do I wonder what it will feel like...some people feel a sense of *deepening*...others experience a change in the feelings...and other don't notice how things have changed...and as you explore this picture I wonder where that chair will be...and when you discover that chair I wonder whether you will notice the sensations associated with approaching that comfortably familiar chair...and as you take a few moments to...relax into the chair...you can get a sense of closing your eyes and drifting off deeper into a dream...'

(Deepener) Pictures in pictures

'That's right...as you now begin to imagine a picture of a beach in front of you...and as you pay close attention to that beach you can begin to notice something different about that picture...and the more you try to see what isn't there the more you find yourself drawn into the image...and then in a blink you can find yourself in that picture searching for a path (pause for about 30 seconds)...and as you wander around wondering where the path will be you can notice it in another picture...and a curious thing about these pictures is that each time you drift into one you go even deeper into a trance...and in a blink you can find yourself on that path...searching for something new (pause for about 30 seconds)...and I wonder what sounds each foot step makes as you walk along that path until you see that new picture...and in a blink you can find yourself in that new picture knowing that what you can learn and discover here can stay in this picture for you to use instinctively in the future like unconscious knowing (pause for about one minute)...'

(Therapeutic Script) Metaphorical 'life changer'

'That's it...and as you continue to *relax deeper and deeper* you can listen to me in the background...and as you listen to me in the background you can understand what I'm saying on an unconscious level...many years ago there was a prince that lived in a castle...and one day he was gazing out of one of the castle windows looking out over the land that his family ruled...and from his perspective he could see people struggling...he could see people suffering...he could see people starving...while he was in this castle getting whatever he wanted with the ring of a bell...the prince ask his father why things should be this way...and was told the world is the way the world is...there's nothing you can do to change it...the prince wanted to know why not...and asked what about all of those poor people...and we are so rich...can't we help them out...his father didn't answer...the prince thought about his life...he had everything money can buy yet he felt unhappy and felt like something was missing from his life...while many of the people in the town around the castle appeared to struggle and suffer yet they would often be smiling and looking happy...that night when everyone went to bed the prince snuck out of the castle in disguise...he had to *discover the answer*...how can people be so happy with such miserable lives...he walked through the town and into the nearby forest...the forest was dark and eerie and the night was still...there was shards of moonlight glistening down through the leaves of the trees onto the forest floor shimmering a path for the prince to follow...the prince followed this unknown path to discovery and wonder and was curious where it would lead him...as he continued to walk the prince could hear noises of animals and birds moving in the dark...he could feel his heart beating loudly as he continued into the unknown...after a while the prince found a clearing and saw a frog sitting by a pond...as he approached the frog it started to talk to him...*'you are on a journey of discovery and wonder and you can wonder what you will discover on that journey'*...the prince began to wonder as he continued to wander through the forest...he thought he could hear something behind him moving in the dark...following him...he could see the foot of a mountain and wondered how he would reach the cave...then out of the dark some tribes people appeared...the prince couldn't understand what they were saying but could tell they were friendly...they began to cut down some trees and build a ladder...the prince decided to help them...after some time they managed to complete the ladder and the prince climbed up to the cave...in the cave was a flickering fire...he sat down by the fire watching the flickering flame and noticing how the light from the fire was dancing on the walls...as he continued to watch the flickering flame he began to see things and his mind began to wonder...then everything went dark for a moment before he found himself resting beneath a tree...and as he rested beneath that tree he began to *discover the true meaning of happiness and life*...and he knew that when you get hot you enter a pool to cool down...and after a while you need to warm up so you go back out in the sun...he then got an image of the Yin Yang symbol...then a coin flipping in the air...then he saw a person struggling and while the person was struggling he saw the person step outside themselves and the part of them with the problem began to change how they were dealing with the situation...then the prince saw people socialising and talking with each other...different images and ideas drifted in and out of the princes mind...and with each image the prince was learning something new about himself that was beginning to change his life in ways he didn't yet know...and the prince continued to notice what seemed like random images until they disappeared when his unconscious knew it found the answers...(pause for about a minute)...and when the time was right he drifted back to that cave before climbing down the ladder and enjoying an adventure of learning and discovery as he headed back to the castle where he knew how the next day he was going to change his world forever...and improve the lives of those in the town...(pause for about 3 minutes)...'

(Trance Termination) Exiting trance now

'That's right...and you can now just take a few moments to allow yourself to get a sense of drifting back into that chair...and I don't know whether you will take 4, 5, 6 or 8 breaths to fully re-orient back to the chair...before you then start to work your way back along the route you took into trance...and as you do your unconscious mind can ensure that all the new re-programming has been installed throughout the mind and body in every cell and neuron...your unconscious mind can honestly and fully integrate all your new learning into that deep and instinctive part of your mind while you continue your journey back to the here and now...and I don't know how much of what is there will be left there as you continue back and how much you will bring back with you...an it's interesting how much you know you can be aware of while you are sat there in that seat...and you can now find yourself working all the way back away from that state to the here and now becoming more aware of sounds around you...of me of what you will be getting on with later and other random thoughts that start to cross the mind and in a moment I'm going to count to 3 and on the count of 3 you can open your eyes and be fully back in the room...one...two...three...and open your eyes...hi now you came here to see how you are able to be helped and I asked you to close your eyes earlier...so do you have any plans for later...(have a brief chat then if you are going to set any tasks etc. you can do that before they go)...'

Hypnotic Inductions, Deepeners & Priming

Induction Introductions

Induction introduction for clients that have been hypnotised before

Prior to starting the induction you will have asked the client about whether they have been in a trance before or not. If they have then you will have gathered some information about this trance experience or their favourite trance experience. Just talking with them about this will have primed the client for entering trance again and they are likely to have already begun to re-enter trance here and now with you whilst you ask them questions and they recall that previous trance experience.

'As you think about a pleasant hypnotic trance experience just allow your eyes to close...and as your eyes close you can begin to recall what...*going into a trance*...feels like...you can recall how you felt before you knew you were going to...*go into a trance*...and...*become absorbed in the experience*...and as you...*recall that trance experience*...you can remember what you could see and hear around you...and while you remember what you can see and hear you can begin to...*experience a little of that trance here and now*...with each breath you take...you can...*become more absorbed in this experience*...and my voice can comfortably go with you as you drift and float deeper and more fully into this experience...preparing to absorb all that is useful and appropriate...and my words can help you to access inner resources and strengths and skills...as...*you hear my voice and relax*...*your unconscious*...can pick up on all the different meanings in my communication...I don't know whether...*you will drift deeper*...with each breath you take or with the sound of my voice or the spaces between my words...'

Induction introduction for clients that haven't been hypnotised before

Prior to starting the induction you will have asked the client whether they have been in a trance before or not. If they haven't you will have explained a little about what to expect and you will have asked them what they think it will be like and what they expect. Anything they have said that they are unlikely to experience you will have clarified for them. Like if they say they think they will be totally unconscious and have no memory of what happened and not hear anything. You can explain it is more like when they are daydreaming. They can still hear everything and are still fully aware. It's just none of that seems so important and they are comfortable continuing to drift and dream in their mind.

'As you...*allow your eyes to comfortably close*...I'll explain a little about what to expect as...*you enter trance*...you already have all the knowledge you need to understand how to...*enter a trance*...you have experience of...*entering a trance*...many times each day...I'm sure you can recall times you have been wandering along and discovered...*your mind begins to wonder*...to thoughts, ideas and dreams...or the experience of waiting in line as...*you start to daydream...now*...as you sit there beginning to drift into a comfortable hypnotic trance you can hear the sound of my voice and have an awareness of different thoughts and ideas that come to mind...and while you listen to the sound of my voice I'm going to talk to you...and just like in everyday situations like when you were in school listening in class and you will have occasionally drifted off into a daydream and later realised you don't know if you remember the whole lesson...many people that...*go into a trance*...remember some things and forget others...and you can remember what is important to consciously remember whilst at the same time forget what is unimportant to recall after this session...so there is no need to try to...*go into a trance*...it can develop all by itself...and you can develop that trance in your own unique way based on your own lifetime full of experiences of entering trance states easily and effortlessly...from all those times you have discovered yourself to have been daydreaming...through to those times...*you become deeply absorbed in what you are doing*...so now...with your eyes closed you can enjoy a moments quiet while...*your unconscious*...prepares to enter a resourceful trance state...that is just right for the work we are doing here today...(pause for about 30 seconds)...that's it...now as people...*enter trance*...they often begin to notice how the breathing changes...for some people the breathing speeds up a little before it settles on a relaxed rhythm...for others it just begins to...*comfortably relax*...and while the breathing helps to deepen and maintain the trance many people become intrigued to discover how much more youthful they look as their facial muscles smooth out...this smoothing and relaxing of the muscles in the face and throughout the body and the continued feeling of comfort that people feel when you come out of trance later are some of the side effects of...*entering a hypnotic trance in this way*...so while you prepare to...*enter this trance state*...I wonder what other positive side effects you will receive...and while you continue into this state...your mind will wander at times and whatever happens my voice can go with you...'

Targeting specific types of client

Logical analyser

'And there will be times when you can focus on what I am saying and pay close attention to the words I am using and all the different meaning that I am conveying in what I am saying and how I am saying it or you could choose to focus on the stages of the process and on how your awareness changes as this experience goes on...'

Daydreamer

'And you can drift and dream and allow your mind to wander...normally in everyday situations people tell you to pay attention...here you can daydream while you listen to me...and only need to follow instructions occasionally as you become more absorbed in the experience...'

Motivated & willing to follow instructions

'And as you follow what I say you can continue to drift more fully and comfortably into the experience and with each step you can move closer to achievement and success...'

Passive – expects the work to be done *to* them

'And as you breathe in...and out...you can wonder how this will work for you...you know many people experiencing hypnosis for the first time imagine that the hypnotist has to do all the work without realising that the hypnotist only does all the work while they are talking and during the session...as they drift deeper into the experience they...begin to discover that each session is a partnership of two minds....the hypnotist works with the unconscious mind while the client follows instructions inside and outside...the therapy session using the conscious mind...this allows the work to be completed both consciously and unconsciously...'

Polarity responder

'And as you hear my voice you don't have to listen to me...or listen to the sounds around you...or pay any attention to any specific thoughts and feelings...and you don't have to follow instructions if you know of a better way of getting the results you came here to achieve...and you don't have to believe this will work for it to get the results you want...as you listen to me you can respond in your own unique way to get the results you want or follow along to any instructions and suggestions...'

Inductions

Journey along a beach

'(Pause for about 20 seconds)...that's right...now...as you listen to me and continue to breathe in that way and notice certain sensations you can continue to become more fully absorbed with the sound of my voice or the spaces between my words while I talk to you...and you know there are many different ways of experiencing the enjoyment of walking along a beach...*as you drift deeper into a trance*...you know some people like to walk bare foot in the sand...feeling that sand on the feet and through the toes...some people like that sand to be dry and warm...others like the sand to be cool and wet...we all have our own preferred experience of walking along a beach...you will have an idea of what you consider to be the perfect beach...what the weather conditions are like...whether it is daytime or night-time...how warm the weather is...what the temperature of the water is...whether you are bare-footed...whether you are walking on stones, shingle or sand and whether the sand is find and powdery or larger grains...or a different shore line entirely...and what sounds are in your experience...whether there are other people nearby or in the distance...or children playing...or the sounds of birds or boats out to see...or other sounds unique to your experience...*as you drift deeper into a comfortable trance* only you know what your experience is that you can internally enjoy...and I don't know whether you will be just listening along and relaxing...or imagining your own experience whilst *you drift deeper into a trance*...or whether you just have a sense of strolling along a beach...(pause for about 20 seconds)...and some people like rough water...others like the water calm...others like it somewhere in between...and some people don't mind any weather as long as they are prepared for it and dressed appropriately...and as you continue to become more fully absorbed with each breath as the waves roll in...and out...(said in time with the clients in and out breath) I wonder whether you can notice a small inviting hut in the distance...a hut that you don't recognise yet it seems familiar...and as you approach that hut you can *drift deeper and deeper* with each step you take...*going deeper* only at the rate and speed that is comfortable for you...and when you arrive at the hut you can be curious about what is inside it...(pause for about 20 seconds)...and when you are ready you can begin to open the door to see a chair you don't recognise yet it looks so familiar and inviting...after getting a sense of your surrounding you can make your way to the chair and relax yourself into it and get a sense of closing your eyes as you *become highly receptive to ideas and concepts that help you to achieve success...drifting deeper with each breath you take...*'

Journey through a forest

(Pause for about 20 seconds)...that's right...now...as you listen to me and continue to breathe in that way and notice certain sensations you can continue to become more fully absorbed with the sound of my voice or the spaces between my words while I talk to you...and as you listen to the sound of my voice you can begin to get a sense of what it feels like to be wandering through a forest...some people find they recall past experiences or imagine a familiar place...others just discover the experience unfolds as they become more absorbed with each breath they take...and as you have a sense of wandering through a forest there are certain sights and sounds that you can begin to become aware of...and I don't know whether you will just *develop a sense of those sights and sounds*...or whether you will *begin to dream about wandering through that forest*...or whether you will just *enjoy drifting into a focused comfortable trance state* as you listen along to what I'm saying...and when you wander through a forest you can feel each step you take...noticing the ground beneath your feet...the sounds associated with each step and how each step can sound different as you notice the different shades of greens and browns and other little flashes of colour...and sometimes you can see beams of light breaking through the leaves shimmering on the forest floor...and feel the air on your face as you hear the sounds of the forest...and it can be surprising to some people to discover just how quickly you can *become absorbed in the experience*...and you know there are some sounds everyone associates with forests in the same way that there are certain sights everyone recognises about forests...and I'm sure you know what those experiences are...what people often get surprised by is how the smells of the forest change as you wander...*deeper and deeper*...in the same way that you can walk through a garden and smell the different plants...and as you continue to wander in your own way *deeper and deeper* I wonder when you will notice the stream in the distance and what you will notice first about that stream...and when you discover the stream you can begin to follow it until you discover a cave that looks so inviting and familiar yet you know you've never seen it before...(pause for about 30 seconds)...and as you enter the cave you can discover a chair you don't recognise that looks familiar and comforting...and begin to feel compelled to...go sit in the chair...and as you sit down deeply into the chair you can begin to wonder what you will experience next...and you can get a sense of closing your eyes and relaxing even deeper into a trance...'

Journey through an art gallery

'(Pause for about 20 seconds)...that's right...now...as you listen to me and continue to breathe in that way and notice certain sensations you can continue to *become more fully absorbed* with the sound of my voice or the spaces between my words while I talk to...*your unconscious...now*...and you can discover yourself entering an art gallery...and as you enter I wonder whether you can notice the way sound changes as you enter and the way the atmosphere seems to *become calmer and more relaxed*...and as you explore the gallery different art work can give you different feelings and different meanings...you may look at a sculpture and understand that it started as just a lump of stone that the artist believed could be more and knew all it needed was to have some time and attention spent on removing what isn't necessary and reshaping what is left...and different pictures will convey different meanings...and as you continue to walk deeper into the art gallery you can wonder which painting you will feel a connection with and feel drawn to...and you may want to get a sense of a number of pictures before finally settling on the one that draws you in...and as *you drift deeper* with each out breath while exploring the various paintings my voice will quieten down in the background for a minute and you can take as long as you want to explore the gallery...and when you next hear me talk you can be prepared to step inside that picture ready to search for a comfortable and familiar chair you've never seen before...(pause for about 1 minute)...that's right...preparing now to step into that picture...and as you do I wonder what it will feel like...some people feel a sense of *deepening*...others experience a change in the feelings...and other don't notice how things have changed...and as you explore this picture I wonder where that chair will be...and when you discover that chair I wonder whether you will notice the sensations associated with approaching that comfortably familiar chair...and as you take a few moments to...relax into the chair...you can get a sense of closing your eyes and drifting off deeper into a dream...'

Journey through a country meadow

'(Pause for about 20 seconds)...that's right...now...as you listen to me and continue to breathe in that way and notice certain sensations you can continue to become more fully absorbed with the sound of my voice or the spaces between my words while I talk to you...and as I talk to you certain memories, ideas and sensations can be recalled...and you can explore those re-experienced thoughts and sensations and wonder how they will assist you in going deeper into this therapeutic trance state...and you know what it is like for someone to enjoy the pleasant experience of walking through a country meadow...being aware of the sights, the sounds, the smells and what it feels like...you have the ground beneath your feet...the sun in the sky...birds flying around and perching in trees and bushes and a sensation of movement as you wonder along...and you know many people that wander through a country meadow wonder and let their mind drift and dream...they gaze round with relaxed eyes and notice the movement of grass and plants...and while wandering through the meadow you can notice certain small animals focused on living their own lives...just respectfully living in the moment...and as the wondering continues so you can continue to drift deeper and deeper into hypnosis...deeper and deeper with each breath you take...and in the distance is a tree...and in that tree is a tree house...and in that tree house is a door...and as you continue to wander through this meadow perhaps hearing that distant stream...you can wonder what you will discover behind that door...and the more curious you become the faster you can walk...and the faster you walk the more absorbed you become...and the more absorbed you become the more curious you can find yourself...now...I don't know whether you will be imagining the experience of walking towards that tree or whether you will just have a sense of the experience of walking towards that tree or whether you will be just listening along with interest whilst your mind occasionally seems to drift off...the important thing is that while I'm talking to you...*your unconscious*...can be listening along and responding honestly and fully to all that I say of importance to helping you to create positive change...and with each breath you take you can find yourself getting closer and closer to that tree...(pause for about 30 seconds)...and as you now reach that tree you can enter the tree house and notice the door...and this door can be an unfamiliar door that somehow seems familiar...and as you notice that door you can prepare to open the door and step inside...but don't step inside just yet...just allow the unconscious mind to be fully prepared for a change...too often people rush in before the unconscious is prepared...(pause for about 20 seconds)...and now gently open the door and step inside to see a chair...this chair when sat in is the most comfortable chair imaginable...so just take some time to notice your surroundings in there before sitting comfortably in that chair and having a sense of closing your eyes while you listen to me...'

Confusion – lifting and moving the arm

'(Pause for about 20 seconds)...that's right...now...as you listen to me and continue to breathe in that way and notice certain sensations you can continue to become more fully absorbed in the experience...and as *you drift deeper into a trance*...in a minute I'm going to reach over and gently lift up the right arm...and as I move that arm you can just *let your mind wonder*...(reach over and gently hold the clients arm just above the wrist. The idea is to move the arm so gently that the client isn't fully aware of how much they are moving their arm and how much you are moving the arm. Then start by saying 'and as *you go deeper into a trance* your arm can go up' - as you say this move the arm gently downwards - 'and your arm can go down' - moving the arm gently upwards - 'and your arm can go left' - moving the arm to the clients right - 'and your arm can go right' - moving the arm to the clients left - 'and that arm can move down...*deeper and deeper*' - moving the arm down - 'and that arm can move left' - moving the arm right - 'and that arm can move up' - moving the arm down - 'and that arm can move down' - moving the arm upwards - 'and that arm can move right' - moving the arm left - 'and that arm can move down...*deeper and deeper*' - moving the arm down further than normal - 'and that arm can move right' - moving the arm to the clients left - 'and that arm can move left' - moving the clients arm right - 'and that arm can move up' - moving the arm down - 'and that arm can move down' - moving the arm up - 'and that arm can move left' - moving the arm right - 'and that arm can move right' - moving the arm left - 'and that arm can move down...*deeper and deeper*' - move the arm down towards the leg - 'and as that arm touches the leg you can begin to get a sense of noticing a familiar comfortable chair you've never seen before... - lowering the arm the rest of the way gently to the leg)...and you can walk to that chair and as you sit down in that chair you can get a sense of closing your eyes and drifting deeper into this trance as if you are drifting off into a wonderful dream...'

Confusion induction

'(Pause for about 20 seconds)...that's right...now...as you listen to me and continue to breathe in that way and notice certain sensations you can continue to become more fully absorbed with the sound of my voice or the spaces between my words and you can wonder whether you truly know which hand is the right hand and which hand is left and whether you would still be right about which hand is left if you had the hand you thought was left put right behind your back and wondered which hand is left in front of you...and the hand that was right becomes the one that's left while the hand you thought was left is actually right behind you...and the hand that is on the right side for me is on the left side for you and the hand that is on the left side for me is on the right side for you...and you can remember to forget what isn't important to remember whilst at the same time you can forget to remember those things that you don't need to remember being aware of thoughts that are left knowing that the right thoughts are forgotten...and you can go deeper into the right state of mind as the conscious part of you is left to think about other thoughts and sensations...and you can imagine a familiar comfortable seat in your mind you've never seen before and take some time to relax down into that chair...and as you relax down into that chair you can get a sense of closing your eyes and going deeper into this trance as you continue to listen to me...'

Early learning set induction

'(Pause for about 20 seconds)...that's right...now...as you listen to me and continue to breathe in that way and notice certain sensations you can continue to become more fully absorbed with the sound of my voice or the spaces between my words while I talk to you...and as you listen to the sound of my voice you can be curious about how we learn to speak...most people have had the experience of learning languages at school and know how difficult that can be...yet without any formal lessons children learn to understand adults and to use words and phrases and learn how the way something is said and the order the words are spoken in all influences the meaning of the communication...in the same way that a child will be born not knowing that it is separate from the rest of the world...then one day the child will be trying to reach for the hand they see in front of their face with the hand in front of their face and discover that the hand they are reaching for is the same as the hand they are reaching with...and as they reach and touch different body parts and different objects they begin to learn what is them and what isn't...they begin to explore themselves as a being...touching a foot with a hand...touching the foot to the mouth...learning...discovering...yet even before this they had already mastered how to breathe...how to move arms and legs and the eyes and the head...they had already mastered the highly complex task of synchronising muscle tension and relaxation to move the muscles in the body...and this was done without any training or lessons they just used that part of themselves that knows more about you than you do...and over time they learn to refine their skill and make more accurate movements and actions...and the unconscious mind is the part of you that whilst you breathe and relax and drift deeper into this trance state can help you to achieve things...it is the part of you that take new learning that you receive consciously and make it instinctive...like when a child learns to walk...they first learn...then they consciously forget...anyone that has tried to walk consciously will know how uncomfortable it feels...in the same way that anyone that has tried to consciously move an arm knows how difficult this is as it happens unconsciously...and as...*you now relax deeper with each breath you take*...you can begin to get a sense of a chair that appears to be familiar yet you know you've never seen it before...and you can get a sense of relaxing deeply into that chair and closing your eyes...and as you do a part of you can draw upon all of those natural resources and skills and the part of you that has enabled you to achieve so much and make so much continued achievement automatic and not even realise how that part of you will do all that it is capable of doing for your benefit and the benefit of others around you...'

Eyes open eyes closed induction

'(Pause for about 20 seconds)...that's right...now...in a moment I'm going to ask you to open your eyes and find a spot in front of you just above eye level to gaze at whilst you keep as still as possible...and whilst you do this your body can *continue to relax* in its own way...and as you look at that spot without blinking I'm going to count down from 3 to 1 and on the count of 1 the eyes can close and you can drift even deeper into that comfortable hypnotic trance state...and whenever I count from 1 to 3 your eyes can open and look back at that spot and each time I count from 3 to 1 your eyes can close and you can *drift deeper into that hypnotic trance state*...and each time you do you can begin to learn how you can *do this for yourself* in the future in situations that are safe and appropriate just by hearing the sound of my voice in your mind counting down from 3 to 1 to...*go deeply into a trance*...and to come back out of a trance you can hear the sound of my voice in your mind counting from 1 to 3...so now just allow your eyes to open and look over at a spot just above eye level...(give them a moment to do this)...that's right...really focus on that spot...focus intensely on taking in as much information as possible...making sounds more crisp and clear (start talking very crisp and clearly)...making everything about that spot sharply in focus...noticing colour...texture...contrast...really paying close attention...focusing all of your attention on that spot...and on the count of one you can *relax that attention* and close the eyes...(count slowly with each count being timed with an out breath and slow enough so that the client is beginning to strain their eyes to keep them open and is really wanting you to reach one)...3...2...1...that's right closing your eyes...as *you relax deeper and deeper*...becoming more comfortable and relaxed...and with your eyes closed the muscles around the eyes can relax...the muscles in the face can relax...and you can *relax deeper and deeper*...and begin to drift off to a pleasant thought or place in your mind...(time each count to opening the eyes with the in breaths)...(pause for about 20 seconds to allow them to begin to drift deeper internally)...that's right...and 1...2...3...opening your eyes...and really focusing all of your attention on that spot...becoming very aware of what you can see...hear...and feel...that's right and (remember to drag out the counting and time it with the out breaths)...3...2...1...that's right closing your eyes...*deeper and deeper*...drifting inside to that pleasant thought or place in your mind...*deeper and deeper*...as the muscles *continue to relax* more fully and the muscles around the eyes can *continue to feel more comfortable* remaining in a relaxed state...as *you drift deeper and deeper* inside...(pause for about 40 seconds)...that's right...and...(remember to time counting up with in breaths)...1...2...3...looking back at that spot...paying it all of your attention...really trying to keep those eyes open and locked on that spot...and 3...2...1...closing those eyes and *relaxing deeper and deeper* into that pleasant experience...that's right...in your own unique way... (Pause for about 1 minute)...that's right...and...1...2...3...trying to focus all of your attention back on that spot...and...3...2...1...that's right...and closing those eyes *deeper and deeper*...as *you drift even deeper* still into that experience...and whilst *you drift deeper* into that experience my voice can continue to go with you as we journey together *deeper into this experience*...and you can begin to get a sense of a chair that appears so comfortable and familiar yet you know you have never seen it before...and when you are ready you can get a sense of relaxing down into that chair and feeling so safe...comfortable and relaxed...you drift off into a deep and dreamy sleep...'

Elman induction

'(Pause for about 20 seconds)...that's right...now...in a moment I'm going to ask you to open your eyes...and when you do I'd like you to prepare to close them again...only this time when you do I'd like you to allow those eyes to close to the point where you know...the muscles around those eyes will be so relaxed...*they just won't work*...and when you know...*those muscles around the eyes just won't work*...I'd like you to test those eyes to...*discover they just won't work*...that's right...and just allow those eyes to open...and now take a deep breath in and as you let that breath out and allow relaxation to occur just allow those eyes to close to the point where...*the muscles around the eyes just won't work*...where the muscles around the eyes become so relaxed...*they just won't work*...and then test to...*make sure they won't work*...(watch for attempted movement in the eyes, or a look of effort on the face. If the eyes open they weren't following along so just say 'that's right now close them again and allow them to relax even deeper to the point *they won't work* and allow yourself to *follow my instructions fully, honestly and completely*...and when you know the muscles around those eyes are so relaxed...*they just won't work*...then test to make sure *they won't work* and allow yourself to go deeper into this relaxed state in your own unique way as those muscles relax more fully...' if they still open their eyes again they clearly are not following along. At this point continue with 'that's right, and each time they open you can *go deeper into this trance state*...closing your eyes now...and open your eyes...*going deeper and deeper*...and close your eyes...(slow down what you are saying to really drag it out and say closing on the out breath, then - for this specific option - say 'going deeper and deeper' on each out breath as well)...and open your eyes...going deeper and deeper...(continue this getting slower and slower and hesitating occasionally until they start struggling to open their eyes) once you see they are struggling to open their eyes or starting to show effort and before they have time to succeed at opening their eyes continue)...that's right...and now just allow those eyes to relax and allow that relaxation to spread down through your body from the top of your head right down to the tips of your toes...and as you do that in a minute I'm going to reach over and I'm going to lift up your right arm and you don't have to do anything other than relax there...I'm going to lift up that arm and let it drop down freely in your lap...and when that arm slumps into your laps you can go twice as deep...(lean over and pick up the arm. It should be limp; the client shouldn't be helping at all and it should drop freely into the clients lap)...(drop the arm) that's right twice as deep...(repeat lifting and dropping the arm and saying 'that's right twice as deep')...and now I'm going to be quiet a moment as you drift even deeper into that state...(pause for about 1 minute)...that's right...and now I'd like you to get a sense of a relaxing chair you've never seen before that looks familiar...and as you approach that chair ready to relax into it...you can be curious about what the experience will be like...and as you relax into that chair you can get a sense of closing your eyes and drifting comfortably and deeply...and my voice can go with you as your mind wonders through different thoughts and ideas and sensations as a part of you listens to the sound of my voice here while another part of you can understand the meaning of my communication...'

Four seasons induction

'(Pause for about 20 seconds)...that's right...now as you listen to me I will talk to you...and as I talk to you your mind can wonder...and while your mind wonders you can begin to think about different things...and as you think about different things certain ideas and feelings can begin to come to mind...and when those ideas and feelings come to mind you can find yourself *drifting deeper and deeper* into the experience in your own unique way...and you know that spring is a time of new life...and you can be out in the countryside and notice how the air seems so fresh and clean...how the colours can appear so bright and vivid and how much variety you can notice...there are sounds you recognise in the spring...like certain sounds of birds and animals...and sights of young animals experiencing the first few months of life...and when a breeze blows it can create movement...and you can notice familiar smells that change depending on where you are and what you are doing...and you are aware of how the state of things can change as summer approaches...and the weather changes and you know what that's like...and the wildlife changes...and the sounds you can hear change...and I wonder what sounds you feel are summer sounds...as you have a sense of what makes summer summer...and there will be certain summer smells...and you know people feel different about different seasons depending on their age and people like different things about the different seasons at different ages...and I wonder what you enjoyed about the summer when you were younger...and as you drift through the seasons certain thoughts and feelings can come to mind...as you drift into autumn...and you know how the colours change in autumn and the weather changes and the smells change...and I wonder what else changes for you in autumn...and many children enjoy crunching the fallen golden brown leaves with each step they take...and we begin to become aware of the nearing winter...and as winter falls so things change again...a scene may look different than it did before...and I wonder what you used to enjoy as a child about the winter period that isn't the same for adults...and imagine what many children feel like when crunching through snow or playing in snow...and the air is often crisp and colour often appears to have been drained out to greys...whites...and blacks...and with little wisps of colours...you can get a sense of noticing a warm and inviting log cabin...and as you approach that cabin it can seem vaguely familiar...yet unknown...and you can enter the cabin and notice what the door was like to open as you see what the inside of that cosy...*relaxing*...warm and comforting cabin looks like as you notice a chair...and you can walk to that chair getting a sense of what it will feel like to *relax* back in that chair and *drift off into a pleasant dreamy sleep*...as *you now relax* into that chair and find yourself instinctively drifting into the most pleasant and wonderful dreamy sleep...and I don't know whether it will be a dreamy sleep from the past or an even more pleasant and wonderful sleep as you continue listening to the sound of my voice...'

Staircase induction

'(Pause for about 20 seconds)...that's right...now as you listen to me I will talk to you...and as I talk to you your mind can wonder...and while your mind wonders you can begin to think about different thoughts and feelings while you *begin to drift deeper inside*...and you can have a sense of noticing a grand oak tree...and you can be curious about that tree as you begin to walk toward it...and as you walk closer and closer to the tree you can notice the secret door in the base of the tree...and you can walk to the door and gently open it to reveal a grand and wonderful staircase with 20 steps...and in a moment you can begin to walk comfortably down that staircase one step at a time as I count down from 20 to 1...and with each count you can go one twentieth of the way *deeper into a trance*...(make each count on the clients out breath)...as you step onto step 20...that's right *deeper and deeper*...and you can notice what is around you and I wonder what you can see...and I wonder if you can hear each step you take and what it feels like to take each step *deeper and deeper into trance* in steps *relaxing into the experience*...19...*deeper and deeper* in your own unique way...and I wonder whether you will take a step with each count or two steps with each count at times and at other time pause for thought or to *relax a little deeper* or just to look around or to stop a moment to *take in the experience*...18...*going deeper and deeper*...knowing you can take yourself down into this state anytime you please just by finding that tree...walking through the door...and walking down the steps...17...*going deeper and deeper* with each count...one twentieth of the way into trance with each number...16...15...14...that's right...and I don't know whether you will *go even deeper* with the words that I say or the spaces between the words...13...12...11...*deeper and deeper*...10...9...over halfway...now into a deep and comfortable hypnotic trance...8...all the way...7...*deeper and deeper*...6...as you approach the bottom of the stair I wonder whether you will become curious about what you will discover...5...as you shortly begin to notice a door down there...4...all the way into trance...3...as you approach the bottom of the steps...2...and 1...in *trance now*...as you push that door and see a familiar looking room you know you've never seen before and wonder where that chair in the room will lead you...and you can...go sit in the chair...and *drift deeply inside* getting a sense of finding that chair so comforting and relaxing your eyes relax shut and you get a sense of drifting off into a pleasant dream...'

Conscious/unconscious separation induction

'(Pause for about 20 seconds)...that's right...now...(move your head slightly left when talking to the conscious mind and slightly right when you talk to the unconscious mind)...as you know you have a (left) conscious mind...and an (right) *unconscious* mind...and when I talk to (right) you (left) you don't have to listen...and there will be times when (right) I will be talking to you and (left) you will just hear me talking...and (right) as you here hear me here and understand what I'm saying (left) you will hear me there and hear me speaking...and (right) the unconscious mind is the part of you that knows more about (left) you than you do (right) *your unconscious* is the part of you that create all of the instinctive responses...that understands patterns and contains a lifetimes worth of knowledge and experience and skills and an ability to use all that wisdom to successfully solve problems when...*the time is right*...(left) whilst the conscious part of you can just *drift and dream and relax* as (right) you do what is necessary...(left) and you can notice a chair that feels so inviting...and as *you relax* into that chair you can get a sense of your eyes closing (right) as *you drift into a dream*...'

Body scan induction

'(Pause for about 20 seconds)...that's right... (Reading slowly and timing sentences to each out breath and leaving plenty of pauses. If the client has been showing some tension follow the script but have the client gently tense muscles first asking them to 'tense the muscles...that's right...and then let that go...that's right...' make sure they have no health problems like arthritis or joint problems etc. if you do this) now...take a few moments to focus on the top of your head...notice how it feels and what sensations you have...allow your attention to drift down to your neck...notice how a little more relaxation can drift in with each breath...and how the muscles can soften as you allow your attention to flow through each part of your body...as you now focus on the tops of your arms...and I wonder which arm the relaxation will spread down first the right arm or the left arm...and as that relaxation continues to flow down the arms you can notice how each breath draws in something new and expels what your mind and body doesn't need...as you focus on those hands...focusing on what they are resting on...noticing which hand is the heaviest and which is the lightest...being aware of different sensations and feelings as you allow the wave of relaxation to spread down your chest towards your stomach...like a healing light cleansing your mind and body...filling your heart and lungs...helping that healing light to spread around your whole body and mind...as that relaxation flows down into your stomach...that's right...*relaxing deeper and deeper* and more fully as a body while the mind becomes comfortably focused...as you allow your attention to drift down into the legs...and I wonder which leg the relaxation will spread fastest into...and as that relaxation drifts all the way into the feet you can be curious how the mind will follow once the body is *fully relaxed*...and now as a mind begin to get a sense of entering into your mind and flowing gently around the body checking that each muscle is *comfortably relaxed*...have a sense of drifting through the head...down into the shoulders...and down into the neck...and now as I remain quiet for a few moments you can take as long as you need to drift through the body filling the body with a healing light and when I start talking again you will have finished and you can get a sense of being in the most wonderfully comfortable chair...as you drift through the body...*now*...(pause for about one minute)...that's right...finding yourself in that chair...*feeling so comfortable*...as you now have a sense of wanting to *drift off deeply asleep into a dream*...and have a sense of closing your eyes while *you become absorbed in that dream*...'

Arm catalepsy induction

'(Pause for about 20 seconds)...that's right...now in a moment I'm going to reach over and gently lift up that right arm...and when I do I'm not going to tell you to put it down...(reach over and gently lift up the right arm. The idea is to be so gentle the client almost can't tell if you are lifting their arm or they are)...that's right... (Start tapping so that you begin to create a state of catalepsy. Tap near the wrist on the forearm. Tap on the bottom of the arm, then top, left then right. Tap at a comfortable pace - not too fast or slow - take the lead from the client and how they are responding. Always tap opposing muscles so when you tap the top then tap the bottom, when you tap the left then tap the right, if you tap the bottom tap the top, if you tap the right tap the left)...and as I continue tapping I wonder whether that arm will seem like your arm but won't feel like your arm or whether that arm will feel apart from you yet you have a sense it's a part of you...that's right...*drifting deeper inside with each breath*...and you know you can notice an arm and not realise it's yours or know it's your but not feel it...that's right...*really absorbing*...that's it...and you can discover something without knowing you've discovered it or know you've discovered it without knowing what it is...that's it...and in a moment I'm not going to tell you to put that arm down any faster than *you drift deeper into a comfortable trance state*...and you don't have to *be fully in a deep and comfortable hypnotic trance* until that arm lowers all the way down to your lap...that's right...(stop tapping and leave the arm 'floating' in the air. As the arm lowers just watch it. If it lowers quickly suggest 'that's right and you can just allow that lowering to comfortably slow down...that's right'. As the arm lower occasionally in time with outbreaths suggest 'that's right' then as the arm finishes lowering fully suggest 'that's it...all the way now...*deeper and deeper into a comfortable hypnotic trance*...') and with that arm resting there you can begin to get a sense of what it feels like to sit in a chair so comfortable you struggle to keep your eyes open...and as you get a sense of those eyes shutting you can drift off into a pleasant and relaxing dream...'

Three things induction

'(Pause for about 20 seconds)...that's right...now...in a moment I'm going to ask you to open your eyes...and when you do I'd like you to just allow those eyes to settle on looking at something in front of you just above eye level...and each time your eyes close here you can drift deeper into a comfortable trance state...so just allow your eyes to open and find something in front of you just above eye level to look at...and as you look at that I'd like to have you hold your focus on that spot and begin to say slowly and out loud three things you can see...(pause as they say three things)...that's right...and now slowly say three things you can hear (pause as they say three things)...that's right...and now slowly and comfortably say three things you can feel (pause as they say three things)...that's right...and now just gently allow the eyes to close...and get a sense of that external scene inside your mind...take three deep breaths as you drift deeper into the experience...and then slowly say three things you can see...(pause as they say three things)...and now three things you can hear (pause as they say three things)...and now three things you can feel...(pause as they say three things)...that's right...and now while you remain as still as you can allow the eyes to open and look back at that spot...(wait for them to look back at the spot)...that's right...and again whilst looking at that spot say three things you can see...(ideally you want them to say different things each time they do this but if they can't think of anything new they can say things they have already said but pause and let them try to come up with new things first)...and now three things you can hear...(pause while they say three things)...and three things you can feel...(pause while they say three things)...that's right...now allow the eyes to close again and drift inside getting a sense of that external scene inside your mind again...that's right...and then slowly take three breaths as you drift deeper into the experience...and then slowly say three things you can see (pause while they say three things - again ideally you want it to be three different things if they can think of three different things)...and three things you can hear...(pause while they say three things)...and three things you can feel...(pause while they say three things)...that's right...each time going deeper and deeper into this experience...and each time finding it harder to open your eyes...and as you continue to relax just allow the eyes to open and look back at that spot...and as you do just begin to say three things you can see...(pause while they say three things)...and three things you can hear...(pause while they say three things)...and three things you can feel...(pause while they say three things)...that's right...and just allow the eyes to close as you drift deeper and deeper...and get a sense of that external scene inside your mind...and as you take three comfortable breaths and allow the muscles around the eyes and face to relax and allow that relaxation to spread through the body in your own unique way you may find it a bit more difficult to say three things you can see, hear and feel but try really hard to say three things you can see, hear and feel and if...it seems too much effort...at that point just allow yourself to relax and drift deeper inside...(pause for them to say three things for seeing, hearing and feeling. If they manage this repeat the process so far of having them open their eyes and saying three things, then closing their eyes and saying three things - you can repeat the wording above again up to this point until they stop saying three things because it is too much effort. If they don't seem to be responding after a few runs through this then after the last run through of the eyes being open suggest 'that's right...and as you now allow the eyes to close and drift deeper I wonder whether you will notice just how deep in a trance you are going...as you listen to the sound of my voice...some people feel consciously as if nothing is different whilst their unconscious mind becomes more responsive and able to respond fully to all that is helpful to creating the desired change so that they can consciously remain aware of the surroundings until they feel safe enough to...*let go*...whilst knowing they will remember everything they will remember when they come out of trance at the end of the session...')...that's right...and as you drift deeper inside in your own unique way I'd like to have you begin to get a sense of a chair that looks so familiar yet you don't recall seeing it before...and as you drift deeper into the experience just get a sense of walking over to the chair and relaxing down into it as you get a sense of feeling so relaxed you just close your eyes and drift off into a deep comfortable and absorbing daydream...'

Self-suggestion induction (good for people that over analyse)

'(Pause for about 20 seconds)...that's right...and as I talk to you I would like to have you repeat what I say to yourself...and when you do I'd like you to change each sentence I say into a sentence directed at yourself...so if I say 'and you can relax deeply to the sound of my voice'...you can say to yourself 'I can relax deeply to the sound of your voice'...so what I would like you to do is to pay close attention and repeat each sentence directing it at yourself...and as you relax there you can hear the sound of my voice (after each sentence pause long enough for the client to analyse the sentence and say it to themselves. They can say it in their head or out loud)...and you can feel sensations in the body...and as you relax deeper you can notice thoughts coming to mind...and you can drift deeper with the spaces between my words whilst becoming more focused with each sentence I say...and you can continue to enter trance in your own unique way...and you know how to walk...and you know how to talk...and you know how to understand what I'm saying...and you know how to let your mind wander and daydream...and you can drift fully into a trance now...and you know how to drift deeper with each suggestion without knowing how you do it...you can just drift deeper effortlessly...and you know what a comfortable chair is like...and you can get a sense of what it feels like to discover yourself relaxing into one now...and as you relax into that chair I wonder when it will become too much effort for you to keep repeating suggestions and instead just let go and drift all the way inside a pleasant dream...all the way...deeper and deeper...and you can get a sense of being so relaxed and sleepy the eyes shut all by themselves as you enter a wonderful dream of excitement and discovery...'

Deepeners

Staircase

'That's right...as you now begin to imagine a secret staircase in your mind...and this secret staircase has 10 steps and with each step you can find yourself going twice as deep into hypnosis...and I wonder whether you will go deeper without realising it or whether you will notice certain changes or sensations...as you step on step 10...going deeper and deeper...stepping on step 9...going twice as deep into a comfortable hypnotic trance state...onto step 8...drifting deeper and deeper...as you step on step 7...deeper and deeper with each breath you take...as you step on step 6...drifting deeper honestly and fully into trance...stepping on step 5...allowing your unconscious mind to find its own way of helping you to go deeper and deeper into trance...stepping on step 4...allowing all of your attention to be absorbed into the trance experience...stepping on step 3...almost there now...becoming deeper and deeper absorbed...stepping on step 2...letting your mind wonder and drift and dream whilst your unconscious focuses on what is important to change...and onto step one...now...into...trance...in your own unique way ready and prepared to make the necessary changes (pause for a minute)...'

Pictures in pictures

'That's right...as you now begin to imagine a picture of a beach in front of you...and as you pay close attention to that beach you can begin to notice something different about that picture...and the more you try to see what isn't there the more you find yourself drawn into the image...and then in a blink you can find yourself in that picture searching for a path (pause for about 30 seconds)...and as you wander around wondering where the path will be you can notice it in another picture...and a curious thing about these pictures is that each time you drift into one you go even deeper into a trance...and in a blink you can find yourself on that path...searching for something new (pause for about 30 seconds)...and I wonder what sounds each foot step makes as you walk along that path until you see that new picture...and in a blink you can find yourself in that new picture knowing that what you can learn and discover here can stay in this picture for you to use instinctively in the future like unconscious knowing (pause for about one minute)...'

Raising and lowering the arm

'In a moment I'm going to lift up your right arm (you can use the left arm) and it can be as loose and limp as a wet cloth...and as I let go of that arm and let it drop freely back down to the leg I don't know whether you will go twice as deeply or four times as deeply into that trance...(lift up the arm (if it feels like they are helping in any way to lift the arm suggest 'that's it just allowing that arm to be as loose and limp as a wet cloth...' and then wait a moment for the arm to become loose and limp. If it doesn't go limp just continue with 'and as that arm lower to the leg you can go deeper and deeper into trance each time the arm touches the leg' and then lift it a few times letting it lower a few times) and let it drop and as it hits the leg say 'that's it deeper and deeper into that trance'. Do this 3 or 4 times. Notice the clients face and body language to notice signs of deeper relaxation like reddening face or slower breathing or slumping more)...that's it...and with each breath you take you can continue to increase receptivity and relaxation and continue to drift deeper and deeper inside...that's right...(pause for about one minute)...'

Down in a lift

'That's right...and I wonder what you would imagine if you were to *imagine a trance lift*...a lift that appears to be in front of you that leads down four floors only as fast as *you drift deeper* into this trance state...how would that lift be decorated...would there be any lift music and if there is what is playing in the background...and I wonder what that lift would feel like as you go down those four floors to the basement of trance...deep into the depths of...*your unconscious*...as you step comfortably into that lift and drift *deeper and deeper* down from floor four to floor three...going only as fast as *you go deeper and deeper*...in your own unique way...down *deeper and deeper*...(pause for about 30 seconds) and you can notice what is at floor three and know it's not the floor for you...as you drift to floor two...and as the lift briefly pauses at floor two I wonder what you can discover...(pause for 30 seconds)...before continuing to *drift down deeper* to the basement...but don't reach the basement yet...allow *your unconscious* mind to be fully prepared for the changes ahead in a way your conscious mind may not know...and the journey to that final level can be a bit further as I quieten down in the background to allow you to complete your journey comfortably and fully...(pause for one minute)...that's right as you reach the basement and you can now exit the lift and enter a room where change can occur...and I wonder whether you will be curious later to wonder how much change was left here with only your unconscious for company after you exit trance...'

Gazing into space

'That's right...you know as a child I used to sit down resting against trees gazing up at the clouds as they drifted gently across the sky...and I used to wonder whether there could be someone out there in space sitting beneath a tree gazing up at the sky watching those clouds wondering whether there could be someone out there in space sitting beneath a tree gazing up at the sky watching those clouds...and I used to wonder what their clouds would look like and what their space would look like from their planet...and I wonder what you think it would feel like to be sat beneath a tree...gazing up into the sky...looking at the clouds wondering whether there is someone out there in space sitting beneath a tree gazing up at the clouds...and I wonder what you think their trees would look like and what their weather would be like and what their clouds would look like...and how effortless it can be to drift across space and time to different planets in the mind as *you drift deeper* into the experience of wondering what would be different for them...what would it feel like to be them...what would be different about the way they think...as they lead a life where they have overcome great adversity and achieved enlightenment and spiritual awakening and explored life with a sense of curiosity and wonder and excitement and achieved great change without even knowing how it happened or how it was possible...(pause for one minute)'

Using silence

'That's right...and in a moment I'm going to be quiet as *you drift even deeper* still honestly and *fully into that trance state*...and you can be curious as to how deep you will go with the spaces between my words and whether you will notice how deep you become as *you drift deeper and deeper*...and I'm going to quieten down a minute and you can take all the time you need to really honestly *become fully absorbed in this trance*...accessing a state of mind ideal for creating change and new discoveries...(go quiet for one minute)...'

Priming

Priming to overcome psychological difficulties

'And in this hypnotic trance state *your unconscious* mind can take on board all that is necessary to help you *see things from a new perspective*...to allow for change to occur...there is an old story told about a monkey living in a forest...the monkey used to worry that one day the sky was going to fall...the monkey used to worry about having an accident and not having anyone around to help...he used to worry about what could go wrong in the future...he used to worry about what other animals thought of him...did he make too much noise...was he attractive enough...he used to worry about times in the past that made him feel guilty...this monkey would create vivid images in his mind as if he was seeing through his own eyes and really feeling all the horrible feelings and worry...one day the monkey heard a crash through the trees behind him...he could feel his heart rush as he instinctively ran to escape the noise...'the sky is falling...the sky is falling'...the monkey shouted as he ran through the forest...the other animals heard this and started to panic...they began running around the forest shouting...'the sky is falling...the sky is falling'...before long all the animals in the forest heard this tale about the sky falling and the forest was full of panic...'the sky is falling...the sky is falling'...just then a wise old owl glided down to talk to the animals and find out what was going on...he spoke to a variety of different animals to discover where this tale came from...eventually he traced the story back to the monkey...the wise owl found the monkey shaking and hiding in the base of a tree...'what's this about the sky falling?'...the owl asked...'I heard it myself...the sky is falling the sky is falling'...replied the monkey...the wise owl decided to take the monkey back to where the sound came from...once they were back where the monkey first heard the sound they waited with anticipation...after a while they heard a crash and a thud...'was that the sound you heard?'...asked the owl...'yes...the sky is falling...the sky is falling'...the monkey replied...the wise owl asked 'how do you know the sky is falling...that sound could be anything...have you checked to see what fell?'...the owl picked up the monkey and flew above the tree tops and just glided around...from up here the monkey could look up and see the sky was still in place and he wondered what had he heard falling...he looked down and watched the trees...just then he saw a coconut fall to the floor...the monkey felt embarrassed but compelled to put things right...he knew what he had to do...he had to let the other animals know it was a mistake...the owl told the monkey he could fly him to the foot of a nearby mountain and give him directions but he had to climb the mountain himself...the monkey understood this...he knew he had a long way to climb so kept his attention on where he was going and how he was going to get there...the monkey climbed and struggled and learnt new things about himself and how much more he is capable of that previously he didn't realise...eventually after many hours of climbing and struggling he felt the satisfaction of reaching the top all by himself...he had managed to keep going even when he didn't think he would be able to...from the top of the mountain the monkey realised he could see the whole forest...the monkey shouted from the top of the mountain and the whole forest listened...as the monkey explain the whole story...and all the animals relaxed...(pause for about 1 minute)...'

Priming to overcome physical difficulties

'And in this hypnotic trance state *your unconscious* mind can begin to make changes throughout the body...*your unconscious* can make all the changes instinctively in the same way that you can learn to walk and talk and adapt over time without any conscious effort at all...*your unconscious* knows how to make changes to muscles...changes to blood flow...changes to the immune system...changes to the digestive system...changes within any cell in the body...you know a master sailor can make sailing appear so effortless...they can tie complex knots...carryout complex tasks...they can keep the boat upright even in strong winds...they learn how to use the wind and the tide rather than fight it...yet when they began they didn't know how to do any of this...some of the skills they learnt through practice...others they developed unconsciously...like their ability to not feel the pain of working their muscles hard...and to be unaware of pain from cuts and bruises and injuries...whilst they are preoccupied with more important tasks...and it can almost feel like they have an out of body experience...their body is doing the work while their mind is observing...and you know once a sailor has learnt a knot and mastered that knot they can tie the knot without paying attention...(pause for about 1 minute)'

Therapeutic Scripts, Processes & Techniques

Smoking habit that currently serves a purpose (like for managing stress or boredom)

'For some time now smoking has been your therapy...and now the time has come for...*your unconscious mind to make a change*...and as you listen to me your unconscious...can discover a new therapy for you...and you know some people smoke to relieve stress before they learn that they can breathe in to the count of 7 and out to the count of 11 and the extended out breath releases endorphins and dopamine and other feel good chemicals into the body and brain leading to relaxation and a euphoric feeling...and they then...*begin to replace the old habit with a new one*...breathing in to the count of 7 and out to the count of 11...other people used to smoke because it was their unconscious minds way of making them take a break or socialise...once they realise this they begin to find new ways to remember to take a break or socialise and as they put these new habits into place the old habit disappears...some people even keep the craving just as a signal...almost like an alarm clock reminding them to take a break and they form the habit of getting the craving which goes once they fulfil it with the new habit...some people find they used smoking as a way of breaking the ice or to meet people...almost like a conversation starter...these people find a new conversation starter or ice breaker that they get in the habit of using before fully stopping the old habit...*and I wonder what it will be for you*...I wonder what your unconscious will find as a new habit for you...and your unconscious can pick up on ideas and possibilities within what I say and the way I say things in the same way that a sprinkle of rain on a seed bed can grow a marvellous array of flowers...and as...*you discover a new personal self-created therapy*...you can begin to...*ditch the old therapy*...and as you listen to this you can...*absorb all that is useful and relevant*...and a part of you can now drift off into the distant future looking back on this present...and as that part of you looks back it can begin to get a sense of how you developed over that time...how you evolved and created your own therapy...how you formed new habits and grew out of old habits...and...*create your own healthy therapy*...you know as a child many people suck their thumbs...many people have imaginary friends...or need to wear nappies...yet many people grow out of these old habits without really thinking about how...*that happens all by itself*...and you can now take some time to begin to create new changes in the mind and body as I quieten into the background for a few minutes to allow that reprocessing and reprogramming to occur on a deep unconscious level...(pause for about 3 minutes)...that's right...*really developing in your own unique way*...and you know sometimes it's nice to move from a smog filled congested city out into the fresh clean and relaxing air of the country with the open roads and clear views...(pause for about 1 minute)...'

Smoking habit that no longer serves a purpose

'That's right...now...I'd like you to take a few moments to get a sense of drifting back into the past and discover a part of you that used to carry out that smoking habit for a purpose...you know many people start smoking and continue to smoke for a reason...over time some reasons just stop existing yet the habit remains...and I would like you to take some time to honestly and genuinely thank that part of you that created the habit...many people find that smoking becomes their therapy...some people continue with the therapy long after the problem has passed...a bit like someone taking pain medication and getting so in the habit of taking that medication they don't stop even when the pain is gone...and you know you can...*stop now*...and you can...*get in the habit of doing something new*...you have a lifetime of experience changing habits without knowing quite how they changed...many children sucked their thumbs yet one day they spent a few hours where they forgot to suck their thumb...and another day they forgot entirely...and over time they forgot more and more until they suddenly look back and realise they hadn't been sucking their thumb for a while and yet they didn't miss it because it happened naturally in its own way and its own time as they no longer needed that old therapy and they grew up...and you know the country with its clean air and open roads can be a wonderful alternative to smoggy congested cities...and I don't know whether you will grow up quickly or take a few weeks to...*really settle into the new you*...and you can...*now take the time*...to...*really begin to undo all those old habits and develop some new healthy habits*...and as...*you do that*...on an unconscious instinctive level I will quieten down in the background and you can take as long as is necessary while I'm quiet to...*make the necessary changes*...(pause for about 3 minutes)...'

Scrambling

Find out from the client what the stages are of the smoking problem for that specific client. For example:

1. Experiencing stress
2. Getting a craving
3. Thinking 'I need a cigarette'
4. Grabbing cigarettes
5. Opening the box
6. Taking a cigarette
7. Putting the box back down
8. Lighting the cigarette
9. Taking the first drag deeply and really savouring it
10. Smoking the rest of the cigarette
11. Savouring getting the last drag out of the cigarette
12. Putting it out

Then get the stages of what the client would like to instead in the same situations. For example:

1. Experiencing stress
2. Recognise I need to take time out of the situation

3. Think of a way out of the situation

4. Go and sit somewhere quiet and do a breathing technique like 7-11 breathing (breathing in to the count of 7 out to the count of 11)

5. Repeat the breathing for about 4 or 5 breaths or until I feel relaxed

6. Plan or write down what I need to do to overcome the stressful situation

7. Put my plan into action

Once you have two lists. One of the problem behaviour and one of the desired behaviour you can begin the scrambling technique. To do the scrambling you want to tell the client that you will guide them through a process based on the lists you both created earlier. You then tell them to imagine 'number 1 of the old smoking habit (and say what number one was – in this case it was 'Experiencing Stress')'. Then to 'nod when they have done this'. Then to 'open their eyes and close their eyes' (as soon as their eyes open tell them to close their eyes) and imagine number 2, then nod the head once they have done this; then open their eyes and close their eyes again. Then number 3...keep repeating this until they have done this for all the stages of the problem. Once they have done that start scrambling the numbers and do this building up speed and as you do it tell them to do each stage quicker and quicker. For example: 'Now imagine number 4 (give the description for this each number) then nod your head, (once they have nodded) open your eyes and close your eyes; now imagine number 9 (description and head nodding), open your eyes and close your eyes; now imagine number 2 (description and head nodding), open your eyes and close your eyes. Repeat this until you have randomly been through all the stages of the old problem behaviour at least four times or until they seem to really be struggling to get the stages in their mind. Then after the last eye closure suggest 'that's right and now just allow yourself to relax and imagine you are...' then read all the stages of the behaviour they would like instead. Build this up as much as possible by asking 'wondering' questions. 'I wonder what it feels like to be recognising you need to take time out of the situation and whether anyone else is noticing that' 'I wonder what others are noticing is different about you as you respond in this new way'.

You don't need to read the numbers out as you read through the preferred behaviour. You want it to flow so that the 'complete response is the preferred one whereas the incomplete 'scrambled' response is the old problem behaviour. This increases the likelihood of the client latching on to the new response as it is complete and easier to follow than the old response.

So with the above example you would read:

'That's right...and get a sense of experiencing stress...and recognise that you need to take time out of the situation...and notice how you find a way out of that situation so that you can go and sit somewhere quiet and do 7-11 breathing...and get a sense of going somewhere quiet sitting down and breathing in counting to 7 and out counting to 11 and notice how that relaxes you...and I wonder where in your body that relaxation will begin and how it will spread...and what others could notice about you that lets them notice that you are responding in a new way...and just get a sense of planning how you will overcome that situation and how you will put that plan into action...'

Then once you have done this 'future pace' by posing questions for the client to think about like 'what will be different responding in this new way' 'who else will notice the change in you' 'what will they notice' 'what other changes may occur because you are responding in a new way'.

Then have the client imagine a variety of situations they would have smoked and to see these situations going in a new way and once they are happy with how they see these situations going then reply those situations only this time step into them to experience the situations going a new way. Noticing what they can see, hear and feel.

Stop smoking script for smokers that have struggled to quit before

'By remaining in a trance and continuing listening to me you will be agreeing to something. You will be agreeing to follow everything I say...and that includes any suggestions for tasks or for developmental opportunities...and it is very important that you understand that by continuing to listen to me you agree honestly and fully to carry out everything I say...so by continuing to remain in this trance you agree to do follow my instructions and suggestions and carry out any tasks...and it's no good just thinking that you are OK agreeing with this but not meaning it because if you do that you will be cheating yourself and you won't get the results that you want...the only way to get the results that you want is to follow exactly what I say...nothing I say will be immoral and everything I say will be for your benefit...but before continuing I need to know you definitely agree 100% to do everything I say...regardless of personal opinions...so you need to be sure in yourself that you agree to that now...and if you do you can keep your eyes closed and continue to drift deeper into this trance...you know people often get people they work with that really annoy them...that really irritate them...they say and do things that really wind them up...and they tolerate them but don't do anything because they think if I say something it might hurt their feelings or they might get angry or it might make it awkward at work...and people have a million and one excuses why they can't say something...and then one day they've been putting up with this for years and years and years and one day they just suddenly decide to say something...they don't think about how they decided to say something just for some reason or another they had that certain mind set where the person said something to them and they just turned and responded...and they found it easy and effortless to do that...they didn't think about it...they just found it easy and effortless to...*just respond*...when the time was right...and they found the result wasn't as bad as they had imagined it would be...and this is a perfectly natural thing...in all human relationships we have situations where people irritate us and we don't say anything because we are polite and kind and eventually we feel we have to say something...and you just say it...and when it reaches that point...almost like the straw that breaks the camel's back...you hold it and hold it and hold it and then it just snaps and you say something and you don't think about the consequences...you just do it...and there was a parent that couldn't get their child out of bed in the morning...then there job was on the line...they were told...if you're not in work you lose your job...and they had a million and one excuses why they couldn't get their child out of bed in the morning...most of the excuses were blaming the child and making out that they were powerless...they would say I have said everything there is to say...I have tried everything there is to try...I have done all the parenting techniques in books...I have done all the parenting techniques on programmes...absolutely nothing works...I have tried everything and nothing at all works...my child still just lays in bed...I can't imagine anything ever working...and the parent would say that over and over and over again...and now their work is saying if you're not in by 9am you lose your job...this parent went up to their child's bedroom and they told their child they are getting up and getting into school because they are not losing their job just because their child is going to be lazy and not wanting to go into school...and they got their child up easily and effortlessly...they got their child into school easily and effortlessly...and they thought it was a fluke...they didn't realise they did anything different...and the next day the same thing happened...they knew the risk to their job if they didn't get their child into school...and they got their child into school...and the day after that...and the week after that...and the month after that...and it was only at this point that they realised their child had turned their behaviour around...and not only had they been putting the blame on the child for why the child wouldn't get up and go to school but they were also putting the success on the child...saying the child must have just grown up out of the blue...they must have grown up and suddenly decided to behave...yet other people could tell that that wasn't the case...that it was actually because they...suddenly had a motivation that meant something to them...and the parent....was now using a different tone of voice...the parent had learnt to be assertive...they had always thought that they were assertive previously...but previously they were going up to their child and saying (without conviction)...but please get up you need to be going to school...now they were going up stairs saying (assertively)...you need to get up you need to go to school...you've got ten minutes...and their tone of voice was totally different...and so the child stopped the old behaviour...and it was because of what the parent did...and we have all had experiences like that where the motivation reaches a point where...*things become effortless and just happen*...there was a person that used to smoke...and they said...I've tried everything I've tried patches...I've tried gum...I've tried inhalators...I've tried changing cigarette brands before quitting...I've tried smoking more and smoking less to see if that would help...and they said...nothing has...*stop smoking*...yet...and then they became pregnant and as soon as they realised they were pregnant they never touched another cigarette...they...*never touch another cigarette*...and that became easy and effortless for them...they didn't think about it at all...they just instantly...*become a non-smoker*...imagine that...you know what it's like...to be doing a behaviour for years and then...*instantly stopping*...and not even missing it or realising or paying attention to the fact that...*you've changed*...because for them when asked...what *stopped* you...you were more than happy to smoke 20 a day...what *stopped* you...they said what *stopped* me was...the second I knew I was pregnant I knew I wasn't going to harm my child or put my child at any unnecessary risk...so I...*stop smoking instantly*...easily and effortlessly...I didn't even have to think about it...and they never had any therapy or anything...the change came from within them...*change just happen*...and we've all had experiences like that...where *change just happens instantly*...and you know the interesting thing about using hypnosis to help people to...*stop smoking*...is that it uses the imagination...and the imagination has tremendous power...some people say they can't imagine things visually...yet on an unconscious level that doesn't really matter because you visualise unconsciously even if you don't consciously realise it...and you know if there was a plank of wood on the floor you could walk across it...and you wouldn't need any safety gear or anything you would just comfortably walk across it...and if the same plank of wood was suspended between two hot air balloons 10,000 feet in the air and you were told without any safety gear to walk between the balloons across the plank you would find it incredibly difficult...and most people would find it impossible...yet you know that you can walk across that plank of wood...because you did it perfectly fine when it was on the ground and it is the same plank of wood...the difference is your imagination is thinking about how you could fall and die or fall and hurt yourself...it's thinking about all the risks...and all the research show that when there is a battle between your imagination and your willpower...*your imagination always wins*...so in that example no matter how much you really want to get across that plank of wood...no matter how much willpower you think you have...you still can't get across the plank of wood...in the same way that when you want to stay awake to watch a film late at night...no matter how much willpower you have to desperately try and stay awake you nearly always just fall asleep...and the more willpower you use...the harder you try to stay awake the more tired and sleepy you feel and the more you just fall asleep...and likewise...when you try desperately hard to fall asleep and you try to put in all of your willpower to force yourself to fall asleep you discover that the harder you try to fall asleep the more awake you stay...and this happens because your imagination is imagining the opposite to your willpower...so when you are desperately trying to fall asleep your imagination is imagining what it is like to be wide awake...and it wins...and when you desperately try to stay awake your imagination is imagining what it is like to fall asleep and it wins...so many people try and quit things using willpower and wonder why willpower seems to not work...and then they say they don't have enough willpower and they use that as an excuse...yet many people that want to give up addictions say they just don't have enough willpower and they don't realise that they could have all the willpower in the world...the difficulty is that the more willpower they use the more the opposite is being portrayed from the imagination making them even more compelled to carry out the old act...and what is needed is to....*stop smoking using the imagination*...not using the willpower...because when...*you stop smoking using the imagination*...when *you quit smoking using your imagination*...then the willpower is trying to have the willpower to start again...and the harder someone tries to start the more they *remain stopped*...and it can be a really interesting situation to be in...where *the harder you try to start something the more stopped you become*...because all of a sudden the imagination is achieving what you want now to achieve...someone I knew lived in a built up area and all the streets were full of congestion...and because of all of the congestion there would often be lots of smog around and noisy disgusting car engines revving up pumping out all the fumes...and you would spend half your time coughing and spluttering and feeling really irritated and wanting to get away from it all...and then moving out to the country meant all of a sudden the roads were open and clear the air was fresh...you could admire the blue sky...at night you could actually see all the stars and it was as if someone had opened up a whole new world of wonder and discovery and you could see things you never used to be able to see and breathe in deep fresh air...and *just enjoy life*...and walk *far more often*...and *savour each and every moment*...someone went travelling once and said they met a tribe in the Amazon rainforest and this tribe told them a story about how there used to be a path they used to all follow...almost religiously they would follow this path...and the path was full of danger...and at the end of the path was treasure...and they wanted the treasure so they would take this path and it was dangerous so not everyone would make it...many people would die trying to make this journey or get severely injured on the trip...and then one day someone came to the tribe and said why don't you just *find a new path*...a safe path...and they had never thought about that before...and this person said if I can find a safe way through as long as you keep following that path and hacking your way through it will stay clear long after I'm gone and you can get to the treasure...so that's what they did...this person hacked through a new path...a safe path...and then they all just *follow this new path*...and they continued to hack away at this new path over and over again making it easier to follow each time the travelled on it...and being in a jungle it didn't take long before the old path had filled itself in and vanished and they couldn't even remember where it was...and now they just had this new path to that treasure...and it wasn't treasure as most westerners would think of it...they thought of it as treasure as it was almost like the gift of life...it was like a pool of water...it was like a spring...so they had fresh water...fresh clean water...but they had got so in the habit before they had this new path of following that old path that it had become tradition to just follow the old path without ever once thinking that it was even a possibility to create a new path...yet once they realised that *you can have a new path*...they just *follow the new path*...and as you listen to me and respond to everything I say on an unconscious level and respond fully and honestly I'm sure a part of you will be curious to know what I was talking about earlier about setting you a task...and at the end I will let you know...and I know that right now you are in a trance...and I don't know how aware you are of the trance you are in...and I know that you are learning....on an unconscious level...how to *create new change*...you know as a child you learnt to count you learnt how to write you learnt how to read and all of these tasks you found terribly difficult to start with...and you really struggled with them...you struggled to notice the difference between a 3 and an m...between a d and a b...between a 2 and a z...there was so much confusion and it was so difficult...yet you mastered this and stopped having to give it any thought in the same way that many,

many years ago you learnt how to stand up and walk and it was so difficult to start with...you would stand up...balancing your weight onto one knee and then the other knee and a hand...perhaps holding onto something and pulling yourself up a bit and as you get yourself up you would swap a knee for a foot and as you put weight on that foot you would fall over...and you would keep trying and trying until eventually you manage to get the second foot up and then you would lift a foot off the ground and you would fall over again...and you would try and try and try again until you got into that position again and this time as you take a step you would move your body weight and discover you can take a step but then when you try to take the second step you discover that if you try to move the same foot again you fall over again...then you keep trying until you figure out you need to move the other foot to take a step and have to shift your body weight to take a step...and it can take many months to master just taking a few simple steps yet once you have mastered it you are growing all the time so you have to keep adapting and changing because your feet are suddenly slightly larger...your legs are suddenly slightly longer...your arms are suddenly slightly longer so your centre of gravity totally shifts so you constantly have to relearn how to walk repeatedly over and over again...but within a couple of years you have mastered this and although you continue to grow and continue to change shape...continue to change size...you continue to *adapt easily and effortlessly* to all the changes that go on around you...*finding it easy and effortless to adapt to change*...now I know that you are listening to me...I know that you are wanting to stop that old behaviour...and I know that you are responding fully to everything I say...and I know that at times you won't even realise that you are doing that...and with your eyes shut and your body just as it is...I'd like you to begin to *use your imagination*...and just imagine...that you go to bed one night...and imagine that you wake up in the morning...and you have amnesia for that old behaviour...you have amnesia for ever having had that old behaviour...so just imagine...*how your day will go*...what will you be doing as you get out of bed...what will the first thing be that you do...and just imagine that...and just imagine running through your day...what will be the second thing you do...how will your whole morning go...and remember as you imagine this...that *you have total 100% amnesia for ever having had that old behaviour*...so it's not a case of saying I wouldn't be doing this or I wouldn't be doing that...the question is...*what are you doing*...so what are you doing in the morning...how are you enjoying your mornings...what things are you doing...what are you having to get done...what are other people saying to you and doing around you...and then as the day progresses what are you doing through the day...really take time to get absorbed deeply...and living a day in your life...the first future day of many...look at things like...when stress occurs in your life...what do you choose to do...how do you...*deal with different stress in your life*...is it that you...*take yourself out of situations*...is it that you...*close your eyes and drift into a deep trance briefly*...is it that *you breathe in a certain way* to *relax yourself*...is it that you hear my voice in your mind...or do you do something totally different or a mixture of these depending on the type of situations...do you keep lists to help time management...are you assertive to make sure you are not overworked...what is it you do...throughout the day...especially at those times where there is extreme amounts of stress that are unavoidable and unexpected...what is it you do so well...really explore that...and take time to explore that...how the way you handle it makes *you feel so good*...make *you feel a sense of pleasure a sense of pride a sense of achievement*...and really explore that...(pause for about 30 seconds)...that's it...and as you explore you way through this day...just explore your way through many other days as well...and even through many different situations and the situations could just come to mind...or they could be contained within future days...in a clear structure...and just allow yourself to explore...what it is like to be in future situations...where perhaps in the past you might have thought...I'm bored...what is it you do instead of being bored...what is it you choose to do...you know the wonderful thing about boredom is it means you've got some free time...and the wonderful thing about free time is that you can do whatever you want for that period of time...because you've got nothing else more important to get on with and if you have then you should be getting on with it...and you know it is easy to become compelled to finish things once they are started...and it is easy to become compelled to start things without any commitment to necessarily finish them as long as it is all positive for your health and wellbeing...so I want you to take plenty of time now to explore many future situations...situations where...perhaps you would have struggled to manage in the past...situations where as *you manage them now*...your brain releases feel good chemicals...like endorphins and dopamine...to make you feel good for making the right decisions and the more right decisions you make the more *you feel good* and the more feel good chemicals get released into your brain...and the more you then *enjoy making right decisions*...and you can take as long as you like to imagine day after day after day cycling round waking up going through a day...all the way through to when you go to bed and fall asleep at night...and then waking up and going through a day all the way through to falling asleep at night...and you can imagine going day after day month after month year after year...some days can just seem to blend into others...being quite mundane with not a lot happening...other days can be full of incredibly challenging difficult events and incidents...that you can enjoy experiencing how *you can overcome those*...in a new way...and just experience year after year after year of what it is like...to live a life entirely without that old habit...because of having amnesia of it when you wake up...and because you have amnesia when you wake up...you can experience what this life is like...and remember...*imagination always out does willpower*...(pause for about 1 minute)...and as you continue to...go through all those different situations in your mind...I'm going to carry on talking to you in the background...and you don't have to pay any attention to me just pay attention to imagining more and more situations that could arise and how they can go...there was a person that used to get bullied...and they would get bullied pushed around beaten up have the mickey taken out of them...they would walk around head hung low...always embarrassed always feeling worthless...and then one day they just thought sod it I've had enough of this I'm *quitting*...I'm not putting up with this anymore...and the bullies knocked this person to the ground...and they just stood up and the bullies knocked them to the ground again and they just stood up again...now previously they used to react...previously they used to either get angry or get scared...either response fed the bully...either response meant the bully could feed off of that response and get more powerful...yet now...the bully would show their face...they would try to get a reaction out of the person...and all the person would do is stand up again...and then stand up again...they didn't show fear...they didn't show anger they showed absolutely no emotion *at all*...and within the space of a day the bully got so bored that they left them alone because suddenly they realised that it didn't matter what they did it didn't work...and the thing about bullies is they often *give up*...if they can't manage to get a reaction whether it's an anger reaction or a fear reaction...they would *just quit*...and they would *stop*...and you know memories are like videos...when you play them back in your mind...they give you certain feelings...but if they are played in fast forward or rewind or with no sound or with the sound sounding funny because they are fast forwarding or rewinding they don't give you those feelings or it could be the picture is too bright or too dark so you can't see what is on the video...in fact you can watch some pretty traumatic stuff on a video and not pick up on what feelings are contained within that video because of the fact that it doesn't look or sound right to give you that type of feeling and your unconscious can understand that on a deep level and apply it where necessary...now I mentioned that there was a task...and the task is this...that if you ever happen to get an old craving...you have a choice...you can either choose to ignore it...in any way that seems fit for you...it could be distracting yourself it could be doing something else...it could be accepting that maybe you have been working for too long a period and the craving isn't actually a craving it is a signal telling you to take a break so you take a break as the action for the craving...or you can carry out the task...and the task is that if you get a craving and you decide that you want to act on it...you are to go outside...and run as fast as you can for five minutes...and time that five minutes exactly so that you run for exactly five minutes not a second faster not a second slower but exactly five minutes...as fast as you can...and then as soon as five minutes is up you decide whether you want a cigarette or not...and if you want one feel free to have one...and if you don't want one feel free not to...so everything is about choice...nothing is taken away from you...everything is about choice and freedom...so if on the off chance you do get any of the old cravings...you can either choose to ignore it in any way you see fit...or choose to accept that it is a signal to tell you that actually you need to take a break because you have been really busy in which case you take a break...but on the off chance you decide you want to have a cigarette then you will feel a compulsion...*you will run*...for exactly five minutes as fast as you can...before then deciding whether to have that cigarette or not...and at that point you can choose not to or you can choose to have it...and you can *always keep in mind*...any time that you have made the decision that you are going to have a cigarette or that you are going to want one...*you will run for exactly five minutes* and at the end of exactly five minutes you will then decide whether you will have that cigarette or not have that cigarette and *that decision will be made at the end of you running five minutes*...(pause for a minute)...'

Once the client has exited trance and before they leave the session set this same task (as mentioned in the script) so that they consciously get told and have a chance to consciously agree to do the task. Tell them you look forward to hearing how they get on next session (if you aren't planning on having another session find a time where you will be calling or catching up to find out how they are getting on).

Quit Smoking Reprogramming Technique

'And as you listen to me I know you want to *quit smoking now*...this process will require your full attention and conscious involvement...what I would like to have you do now is to get a sense of that craving you have had in the past that had led to you smoking...once you have a sense of that craving and can really begin to feel it just let your head nod (wait for the head nod)...that's right...and now get a sense of where in your body that feeling is...as if it is a physical entity somewhere inside your body...get a sense of where it is inside your body...and now whilst you have got that craving there don't let it go just keep it there...and as you do just think of the worst negative effect of giving into that craving that could happen...so it could be something like dying because of ill health like getting lung cancer...or making a child ill because of your smoking...or it could be that you have no money because you are spending it all on smoking...whatever it is for you that you can think of...it doesn't matter if that outcome has never convinced you previously to....*stop*...because many smokers already know all of these outcomes...packets of cigarettes nowadays normally have disgusting pictures on them with the effects of smoking but it hadn't put smokers off but it is only a small part of this process...now...very assertively...say in your mind...*NO*...very assertively say in your mind...*NO*...*you don't want that*...and then when you say that see an image in your mind...almost like a movie in your mind seeing through your eyes hearing through your ears of everything playing out perfectly of having a future with the ideal outcome of saying NO...*what are the benefits of saying no to the old smoking habit (say with curiosity)*...really think about what those benefits are...because anyone that wants to...*quit smoking*...has to have an excellent reason for wanting to...*quit*...it's no good *quitting* for somebody else nobody quits unless they've got a good reason...so it could be for your health...it could be financial...it could be that *you find it very tedious* and you wish you weren't addicted...whatever it happens to be just imagine what you would see and hear and feel having said NO and *feel incredible pleasure with yourself for saying NO*...now open your eyes and then close your eyes again and get that craving back again...try and make that craving as strong as you can...try and increase that feeling as strong as you can...and as you increase that feeling...increase the sensations of the negative side effects of giving into that feeling...and then assertively say NO...and then see all the positive effects and feel incredible pleasure with yourself for winning for saying NO for being a success...for achieving things...and then open your eyes and then close your eyes again and try really hard to build that craving up again really try and build that craving up again and as that craving builds up really build up all those images all those negative images associated with what the outcome is if you had given in to that craving and then assertively say NO...and then see all the pleasure and experience all the pleasure of overcoming all that...experience the pleasure of success...really be aware of all the positive outcomes in the future because of having said NO...and then open your eyes and then close your eyes again...*it will be harder each time*...but really try hard to get that craving really try to increase that craving...really try to increase it...just get all those images again all the negative effects of if you had given in to that old craving...and then assertively say NO...and then get that feeling of pleasure and that positive images in your mind...images of success...images of what life is like...being fitter and healthier...more full of energy...keeping those around you fitter and healthier...what others will notice...what friends will notice...what family will notice...how it will make the home perhaps smell different if you smoke at home...how it will make you smell different...whatever it is for you that comes up in mind...then open your eyes and close your eyes and try and build that craving up whilst building up those negative images...and whilst you build up those negative images then say NO...then get all that positive feelings and positive images...then open your eyes then close your eyes...try and build up that craving again associated with those images then say NO then go to all the positive feelings again and the positive images...then open your eyes and close your eyes...and try and build up that craving again...really try hard to build up that craving again...with all those negative images and then say NO and then see all the positive feelings and positive images and experience all that positivity of having said NO...and then open your eyes and close your eyes...really try hard again to build up that old craving really try hard to build up that old craving with those images and then say NO and then see all the positive images and positive feelings and enjoy that for a moment and then open your eyes and close your eyes...try and get that craving back again...try really hard to get that craving back...with those negative feeling and negative images and then say NO...and then see the positive feelings and positive images and build all that up and let it increase and double that and double it and double it some more...*really cranking the positivity up*...and then allow yourself to *enjoy drifting into the future* to what the future holds leaving here a *new you*...explore what the future hold over the next year and in years to come...what changes will you be around to see and enjoy...what will others notice about this new you...how will you be able to enrich the lives of others because of changes you make here today...just take as long as you need to *explore that now*...(pause for about three minutes)...'

Dissociation and reintegration quit smoking process

'Just imagine a TV in front of you...it can be a big giant plasma screen TV so that you can see everything clearly with surround sound and you've got the remote control so you are completely in control of what is going on...and just get a sense of seeing yourself on that screen...seeing yourself sitting where you are now and I don't know whether the camera angle is from behind you or from where I'm sitting or from in front of you but just see yourself on that screen sitting there...and then just watch as that you stands up leaving something behind...stands up as a non-smoker...now you don't know yet how that you became a non-smoker but that isn't important but just watch as they stand up and they leave here as a non-smoker...they walk through the door and just have that camera follow them as they leave here...and that you goes and carries on doing what you are going to do after this and you go on into the future seeing that you doing what you will be doing later on...and maybe it is at a time you see them looking in a mirror or maybe it is whilst they are having a drink or some other situation you suddenly see a little smile come across their face and you know that although you didn't hear it you know that what just ran through their mind was that they still hadn't had a cigarette and they hadn't even thought about it...and even in situations where they would have done they still haven't thought about having one...that's it...and just imagine watching that you enjoying the rest of the day and evening and then watch that you mixing with other people...smokers and non-smokers...and see that you on that screen really be challenged...perhaps someone is constantly pestering you to have a cigarette...and see your response to them...see that new response...and then watch as that you is just mixing with smokers in a new way...and watch as that you gets into a situation where they feel a little bit nervous...and see how they handle that situation differently...notice how that you can become absorbed in small things when it is the right time to do so...like being absorbed in a certain feeling or sensation or something you can see or hear...and then just move forward slightly further into the future...and watch yourself talking to someone about how pleased you are with your progress...and notice how that you quickly gets distracted to talking about other mundane things forgetting about what you were talking about...and watch that you falling asleep and then waking up the next day and going through a new day in a new way...experiencing various events and situations...watch as that you goes through a variety of different days...watching how that you takes breaks when they are needed appropriately and in a healthy manner...enjoying socialising...eating meals...chatting with people and communicating in a variety of ways...all in a new way...and after a few months see that you talking with someone about how they still haven't had any cigarettes since that day they were hypnotised...and you know there is a voice that you listen to on one of your shoulders that tells you all those things that are good and that are positive and I wonder which shoulder it is on...and that voice has the power to give you pleasurable experiences to remind you why you should listen to it...and you have control of that pleasurable voice and the voice on the other side...and you can turn down the volume of the voice on the other side and you can change the voice speed so that if it tries to tell you to have a cigarette it just sounds like Mickey Mouse and *you just can't take that voice seriously*...and the more that voice talks the quieter it gets...and the quieter it gets the stronger the other voice becomes...and just get a sense of restarting that video again on that screen in front of you and this time step inside that video in to that you on the screen so that it is all in 3D so that you can see what you would see...hear what you would hear and feel what you would feel...get a sense of sitting there and getting up and leaving something behind...and then run through this experience...what you will be doing later...having that small smile as you realise you still haven't had a cigarette...and falling asleep...and waking up in a new day and enjoying it in a new way...and taking breaks...socialising...communicating with others...all in a new way...and notice that at times *when you need to relax your breathing changes and relaxes* and gets deeper and each out breath gets slightly longer than each in breath...that's it...and then just notice yourself in the future...a long way into the future looking back and talking to a young child and the child looks familiar but you don't know why yet...and the young child is asking about that thing that people used to do in the olden days...and that child asks did you ever used to do that smoking thing...and you explain that you did it for a while and that it was a horrible habit and *you stopped*...and like all children they can't stop asking questions and they ask why...and you explain to them why...and they ask other questions...and you tell them the about the day you stopped...and you tell them that occasionally there were times where you would find situations difficult so you would *go and take time out to relax*...and as you talk to this young child you feel a sense of pride of what you have achieved as you look into this familiar child's eyes...and by *stopping smoking* you have had a lifetime to share so much love...and from that one event where *you stopped smoking* there have been many knock on event with positive changes and positive effects on the lives of others and even on the lives of people you don't yet know...and the child asks you what would life be like had you not *stopped smoking*...and as you think about this tears come to your eyes...as you realise that the child wouldn't be there if you hadn't stopped and that you may not have been there...and as you think about what might have been had you made the wrong decision...you get a thought of lying in a hospital bed with your loved ones around you seeing you lying there with the sounds of the hospital equipment and a child crying by the bed wondering what is happening to you and whether you will be coming home...and notice how that thought is the straw that breaks the camel's back...it is the thought that makes you think...never again...and I don't know whether this is the thought that comes to mind whenever you are offered a cigarette or tempted to give in...and the thought that reminds you why you are staying stopped...and that positive voice helps you and encourages you...and just take a moment to see a whole chain of events all happening almost like they are on individual slides with each one different smoking experiences from the past...just watch them all running backwards almost like they are all videos rewinding so that each and every one of those past events of smoking all at the same time are running in reverse so they all start at the end and run all the way back to the beginning...each and every event all the way back to that first ever cigarette you had all run in reverse...with the images happening in reverse...the audio in reverse and the feelings happening in a new direction...and when you sleep tonight your unconscious can carry on this reprogramming...and all those videos can wear themselves out running in reverse then restarting at the end and running in reverse again...then restarting at the end again and continuing this until it just becomes too much effort...and when you exit trance you can bring back a strong sense of *never again* and drift back to the you sat here with me now reintegrating fully and completely into the here and now...(pause for about one minute)...'

Weight loss – trance-like eating, can't stop eating

'That's right...and you know there are times when you appear to be in your own world while you eat...and you can discover when to apply the brakes...and as you listen to me a part of you can begin to explore possibilities and past learning's and even one off events where you instinctively applied the brakes at the correct time without even realising you did it on a conscious level...and you know there have been times when you have eaten a healthy amount and felt full and times when you have pushed just past that healthy amount and felt bloated...and at these times you have a habit of stopping without even thinking about it...and you can learn now on an unconscious level how to apply this to all of your future eating situations...and I wonder when you are in the future looking back what the main thing was that changed an old eating habit...and at what point you noticed eating had become a conscious act and after about 5 minutes of eating it starts to become a chore...you know sometimes the things we love most when done too much become tedious...and you can take some time now to really fully and honestly integrate all that is relevant to making the necessary changes...(pause for about 1 minutes)...'

Weight loss – metaphors for putting in hard work, getting a reward, finding the inner beauty, achieving your preferred physical image

'And many years ago people used to travel hundreds of miles to scoop stones out of streams...and they would work hard to remove as many of those nuggets of Gold as they could...and they knew that all they had to do was work hard...remove the stones and they would be rewarded handsomely...and the job was hard work but they kept the goal in sight...they knew that hard work was for their future...and you know a sculptor can take an ugly piece of rock and they will look at that rock and see the inner beauty...and while they hold that inner beauty in mind they will begin chipping away...removing all the unwanted stone...and as stones fall off...the sculpture begins to take shape until after lots of hard work and sweat a beautiful work of art is created...and you can take some time now to really fully and honestly integrate all that is relevant to making the necessary changes...(pause for about 3 minutes)...'

Weight loss – comfort eating or emotional eating

'That's right...and you can imagine a TV in your mind and on that TV you can see a version of yourself drifting back to a time before any of the emotional or comfort eating began...and you may not even consciously realise when this was...and a part of you can go through all the relevant memories from that point forwards that led to the old emotional eating habit forming and as that part of you works through those old memories it can look for any hidden emotion...you know many people with emotional eating problems discover that when they think back to the events that cause the emotional eating they remember emotions like guilt and shame...and when they focus on those memories a little longer they realise that they also had a feeling of pleasure or control...and your unconscious mind can become aware of these hidden emotions in the old memories and begin to learn to *leave the past in the past* whilst *learning what is important for you* health and wellbeing in the present and future as *you form new and exciting habits*...and while the unconscious mind is forming these new habits your conscious mind can be curious what changes it will notice first...and consciously you can get a sense of different past situations that stand out and imagine them running backwards and forwards rapidly in your mind with all the action happening rapidly...with the speech sounding all sped up and playing backwards and forwards and just notice how the emotion can begin to drain away the faster and faster those old memories flash backwards and forwards to the point where they are just a blur or a flash of light...and after a few moments they can begin to settle back down looking similar but seeming so different...and in a moment I will quieten down in the background while your unconscious makes all the necessary changes to create new coping strategies and opportunities for achievement...(pause for about 1 minutes)...'

Weight loss – limited tastes; eating too much 'junk food'

'That's right...and you know when a child is born they don't know what foods they will like or dislike...and over time they get made to try one thing...then another...and another...some they like some they hate but they *try them all anyway*...and as they grow up some children get used to eating certain foods and it's as if they get stuck in a rut...and as time goes on they forget about other foods they used to like and other foods they do like and stick to what is familiar...and some foods they really didn't like but when they grew up and tried them they found their tastes had changed...and *your tastes change automatically*...and you know I remember this man that used to eat the same meal whenever he went out...it didn't matter what was on the menu he always ate what was familiar to him...then one day he was in an unfamiliar situation out to dinner and the meal he normally picked wasn't on the menu...he didn't want to lose face in front of the couple that had invited him out so he decided to *just pick something different and eat it anyway*...he ate the meal and was surprised to find he actually liked it...following that experience he began to sample other foods on different menus...he discovered he liked more than he realised...some foods he liked a lot...other foods he didn't mind but wouldn't buy them himself if he had any other option and some foods he really didn't like the taste of...and you know the beauty of a rainbow is only possible because it is made up of a full range of colours...and you can *explore new options*...and you know in a mirror everything is reversed and you can see yourself in a mirror and as you step into the mirror and look back you can notice how you appear to be changing for the better over time into the you you are becoming...and you can take some time now to really begin to...*make all these changes*...as you *drift deeper and deeper*...and you know if you wear glasses at first you notice them but over time you stop feeling them on your face...in the same way that you can walk into a room and notice a smell but after a few minutes you stop being aware of it...it's as if the smell is gone...(pause for about 3 minutes)...'

Weight loss – wanting to exercise more

'That's right...and you know people are often surprised about how much exercise they actually get in an average day...and how much exercise you can get just from walking a little faster...*taking a few more steps*...having a leg instinctively *fidget* under a table during meals...when alone...or just when it is appropriate to do so...and it is often surprising to discover just how easy it is to exercise without even trying...and I wonder whether you will find yourself exercising more throughout the whole day or just during specific times of the day...and you can be curious to discover how easy it is to form a new and healthy habit...and I don't know whether that habit will develop into making time specifically to exercise or whether you will just find yourself being more active throughout each day...(pause for a minute)' (Recommended you also use the weight loss metaphors)

Depression caused by issues in the past

'That's it and there are unresolved issues in the past...and you can drift back to a time before any of those events occurred and as you then drift forwards you can get a sense of all of those issues forming slides in your mind almost like slides of movie film...and in a moment you can begin to watch yourself watching those slides like one long continuous movie...and as you watch yourself watching that movie you can notice how it all seems to be in black and white and grainy...and you can watch yourself watching that old movie as it fast forwards and rewinds so fast it looks like a blur...and it can continue to fast forward and rewind so fast any negative emotion just seems to drain freely out of all those old memories...and as it continues to fast forward and rewind certain memories can have bits added...you know we all wish certain events went differently or that we were able to say or do things we never did...and outside of conscious awareness these things can be done or said in a way that resolves those memories *putting them firmly in the past*...and you can take as long as you need over the next 2 minutes to work through all of those memories as you watch that you watching that old movie...*now*...(pause for 2 minutes)...that's right...as you now allow the past to be left right back in the past while focusing on what you have to look forward to in the future...and you don't even have to know consciously what you have to look forward to you can just allow *your unconscious* to guide you towards *pleasure and happiness* while you focus on making sure each foot is pointing in the right direction as you take each step and make sure that you only step with one foot at a time...(pause for 2 minutes)...'

Depression caused by issues in the present

'That's it...and you know juggling in the dark is difficult...you need someone to help by turning a light on...and as you think now about all that is going on in your life you can begin to slow things down a little and take a step back in your mind...and you can step back in your mind and see things from a new perspective...and from this new perspective you can begin to look at what you need to get done and how...*you can do it*...and you can find that place in your mind where...*you can remain calm*...work through all those issues that in your normal waking life was consuming your mind...and you can consume a new way of dealing with things...and you know it is interesting that whatever one person finds a challenge or a problem someone else in the world will find...*it won't faze you*...and you know everyone responds to things differently...almost everyone knows someone either personally or has read about someone or seen them on TV that is in a similar situation yet seems to be *coping fine with it*...and you can be curious about how this is possible and what is different about how they respond...and what *you can learn unconsciously from this*...and how *this knowledge can empower and enhance your future*...and you can take *time now to really begin to integrate all the necessary changes*...(pause for about 2 minutes)...'

Depression caused by worrying about the future

'That's it...now...some people find that they spend their time focusing on what might happen rather than on what does happen or on 'what if's'...and the interesting thing about the future is that it hasn't happened yet...so unlike the past or the present the future can become whatever you make it...and sometimes this can seem unlikely until...*you begin to focus on what you want in the future*...and...*begin to focus on how you will achieve that*...and...*focus on what you need to do to overcome future difficulties*...and...*over time things improve*...and you know at the top of any hill can be a wonderful view followed by a way down the other side...and it's interesting to think that any problem any one person finds difficult to handle there will be someone else that handles that fine...and if you were going to focus on what you want in the future I wonder what it will be...many people think about a question like that and initially respond by thinking about what they hope it won't be...they think...'I wouldn't have money difficulties'...or...'I won't keep getting upset about things'...and this thinking helps to maintain problems...as they...*learn this*...they begin to think about what they actually want and what life will be like as if they were following themselves around with a camcorder...they...start saying things to yourself like 'I have stepped back and sorted out my finances...I have arranged payment plans...I have sorted out my spending and now spend only what I can afford each month...I have got help that I need...I have looked at what used to upset me and worked out how to handle it...I am planning more...when I notice something that used to make me worry I am looking at what I need to do and planning how I will handle those situations...If I can't change something or have no control over something I work out what I can do so that I can handle it remaining calm...if something difficult happens it doesn't help to also be worrying about it...I now make sure I have a clear head instead...I *smile more*...I move around more and *feel more full of energy*...my sleep has improved...people around me seem happier...I *feel more relaxed*...'...and I wonder what changes will be for you and how quickly...*you will begin to notice the changes occurring*...and as a part of...*you give some therapeutic thought to what I have said*...another part of you can begin to...*learn to relax*...and it can take all the life experience you have of relaxing from falling asleep at night to being absorbed in a good book or film or conversation...and can *apply this natural learning to the future*...and as you do that I will quieten down in the background...(pause for about 2 minutes)...'

Depression where the client can't work out why they feel depressed

'That's it...and some people find they can feel depressed without even knowing why...and they can be checked out to and discover there is no medical physiological reason why...yet they still often feel low...and many of these people are often surprised to discover how much emotional thinking they do...and surprised to discover that even excessive positive emotional thinking that is unresolved like really wanting to achieve something and thinking about that thing many hours a day yet never quite managing to achieve it...can lead to the brain feeling overloaded with constantly firing off new patterns like constantly opening new documents on a computer...and eventually it starts to run slower and slower until it begins to crash...and all that is needed is to close many of those documents and the computer works fine again...and *your unconscious*...can understand and know what it needs to do...and while *it does that*...a part of you can begin to notice a large painting in front of you...and you can begin to get your hands in the paint and move all the colours around...mixing colours...blending...and feeling compelled to continue to move those hands around moving that paint around until it feels like the picture is done...and the picture can end up looking like something or nothing...it can be abstract or realistic...you can have a sense of its meaning or have no conscious sense of the meaning of the painting...and as you move that paint around...*your unconscious*...can begin to reorganise connections and pathways in the brain...you know learning and change can happen quickly...everybody has had experiences where they have learnt a lesson from a single event and kept that new way of responding for the rest of their life...that's it...as you now take time to continue moving that paint around until that painting is complete I will quieten down in the background for a couple of minutes and you can take as long as you need over that time to make all the necessary changes and to complete that painting...and you can wonder how life will be different and what changes you will notice in the future and who else will notice a difference and how they will react and what they will say...(pause for about 2 minutes)...'

Worrying about events

'That's it...and you know certain thoughts about certain situations can give us feelings of anxiety nervousness or panic...and other thoughts of situations can help us to *relax* and...*feel calm*...and anxiety is a natural survival response...and in the right place at the right time it can be a useful response...and as *you continue to relax* a part of you can begin to unpick what situations anxiety is useful for and what situations should you respond differently...and you can...*respond differently now*...and many people wonder how...*you stop panic attacks or anxiety*...and to...*stop them before they occur*...you can...*learn to stop worrying*...and instead...*start problem solving*...and the more you problem solve what you used to worry about the more...*anxiety stops*...and you can preoccupy your mind and become absorbed in healthy activities and relationships and conversations...and as you do...*changes will occur in your life*...and people often wonder how...*you remain calm*...and it's useful to remember that a flood of emotion only lasts a few minutes and has to keep being topped up if it is going to stay...and you can rate this emotion in your mind from 1-10 and notice how it changes over time and how quickly it subsides...and notice that any response to that emotion also passes...and during negative emotions you can...*step outside yourself*...and...*become an observer*...while that you calms down...and I don't know how...*you will integrate this learning*...and how...*you will learn more for yourself*...and I will quieten down in the background while...*you take this time to learn and update the mind and body on an unconscious level*...and you know if a car alarm is set too sensitively it keeps going off and annoying people and irritating the owner...if the alarm isn't set sensitive enough someone may steal the car...and just like Goldilocks discovered it has to be just right...(pause for about 2 minutes)...'

Certain situations trigger panic

'That's right...and you can now just begin to get a sense of a TV set in your mind...and on that TV set can be images of situations that in the past caused you to panic when you were in them...and as you look at those still images they can begin to change...and the changes can begin to drain all the emotion from them...then as you continue to watch those images you can be curious to discover what will change as all the images appear on screen like a grid of images...all very small...like thumbnail pictures...and they all begin to rapidly fast-forward and rewind to the point where each picture just looks like a blur...and you don't have to consciously know what each image is of...you can just watch that therapy being done over there on that screen with those images...and as the therapy completes itself the images can one at a time slow back down again and come to a rest...and when they do you can notice how comfortable you can feel as you look at those images and...feel comfortable with those situations...and a deep and instinctive part of you can learn from this experience in a way that is healthy and beneficial to you...as I quieten down in the background while...that work gets completed...(pause for about 3 minutes)...'

Managing pain of an injury that still needs looking after

'That's it...and while you remain totally stationary I'd like to have you get a sense of a healing light surrounding your body...and that light can begin to enter the body with each breath that you take and as it enters your body it can begin to work through the body to that discomfort...and as it reaches that discomfort it can begin to cushion it and wrap around it...and it can cushion and wrap around more and more with each breath that you take...and that healing light can begin to take hold of that old discomfort shrinking it down smaller and smaller...and each photon of healing light is full of energy and the more you keep breathing the more healing light wraps tighter and tighter around that discomfort...and while the unconscious mind continues to wrap that up into an ever decreasing size you can listen to me...and you know you have a lifetime of experiences of treating discomfort...and it's interesting how 70% of discomfort is comfort...and you have experienced playing or having fun or being absorbed in an activity to the point where you don't notice aches and pains or cuts and bruises...and you have a lifetime of experience being so deeply asleep you lose the feeling in your body...and a lifetime of experiences being distracted by something more important...and I wonder what would happen to an experience of discomfort if a hungry Lion started chasing you...and you know people...*forget about discomfort*...and people...*step outside yourself*...and people...*imagine the sensation is different*...and people *habituate to constant feelings* so they...*stop noticing it*...and that healing light really can now wrap into a smaller and smaller ball until it is so small it can fit on a pin head...and as a ball small enough to fit on a pin head just glistening with that healing light it can begin to be moved around a little before you prepare to move it outside yourself so that you can see it in front of you...and when you see it in front of you you can be prepared to...*let it go*...and...let it fall to the floor and roll away...and I wonder whether you will...*feel different*...in a minute or when that ball rolls away...or a few moments after the ball has rolled away...or *just before you open your eyes*...or just before you prepare to come back out of trance or just after those eyes open or before you leave here today...and you can wonder when that will happen...and now over the next 2 minutes you can take as long as is necessary to achieve the comfort you desire...(pause for about 2 minutes)...'

Managing chronic pain – like arthritis

'That's it...and as you listen to the sound of my voice you can begin to *learn how you can remove discomfort in the future*...and you know you are sitting there...and you have discomfort there...and you know that discomfort isn't over here...and it isn't in my chair...and it isn't in the space between this chair and that chair and it isn't on a table or at a table...or in any of the vehicles outside...and it isn't in someone else's house...and it isn't in your bed at home or any other bed...and you know all that discomfort is only there in that chair where you currently sit...now...the other day I was getting ready to go out when someone started talking to me...I put my keys down a moment on the chair while I was listening and getting my shoes on...when I left I discovered I had forgotten my keys...and you know we all have experiences like this...and these experiences seem to just happen...we don't need to try to make them happen...*it just happens all by itself*...and when I was younger I used to wonder why adults didn't take their hair home from the hairdressers...they would go into the hairdressers and look in a mirror and know what they want removed and the hairdresser would be friendly and talkative and would cut off all the hair the customer doesn't want and then the customer would look in the mirror and be pleased and would then pay the hairdresser and would then leave...leaving an old part of themselves in the hairdressers...and I used to wonder why if it was a part of them they didn't want to keep it...one day I asked an adult this question and they told me that...once the hairdresser has removed what you don't want...*you are pleased with the results*...and...you don't want to take what's been removed away with you...you want to *leave it behind*...and so that's what you do...*you have no need for it anymore*...so why would you want to keep it...I understood this and knew it was a bit like how an artist may make a sculpture out of wood or stone and they would carve or chip away all that they don't want...and this can never be reattached...and they take the finished item from the work area to put on display...and the old wood or stone that has been removed can never be put back together with the finished sculpture...and you know that discomfort only exists where you sit...*it doesn't exist anywhere else*...if someone searched anywhere outside of where you are sitting they wouldn't find that discomfort...and you know you can *learn on an unconscious instinctive level*...and you can know how to *leave something behind*...and you know there is a difference between feeling and doing...and you can feel like doing something but not do it...or do something but not feel it...and I don't know whether you will feel like leaving here at the end of the session or whether you will actually decide to leave here at the end of the session...and I know you came to see me because you want to leave something behind...and I know that when you go you will have a feeling you want to leave...and so many people leave their coat on the chair without realising as they leave...and they come back for it...and you can *leave something on the chair you decide you can live without*...and you can learn more from this on an unconscious level than you ever thought possible on a conscious level...and you can take some time now to integrate and absorb all that is necessary to leave here having achieved what you came here to achieve...and you can learn how you can apply what you have learnt here into everyday life and you know it is OK for parts of you to remain in trance with each breath you take will the rest of you enjoys getting on with life...(pause for about 3 minutes)...'

Removing phantom limb pain

'That's it...and I wonder whether you have thought about how by having the skill of being able to create phantom pain you also have the skill to create phantom pleasure...and many people learn unconsciously how to create phantom pain before they...*discover how to create phantom pleasure*...and this is something you don't know how to do consciously...(pause for about a minute)...and I know someone that moved into a home beside a railway line...and those first few night the person found they couldn't sleep...each time a train went by they would suddenly be aware of it...they were finding this to be such a pain...then one day they had a friend round...this friend was complaining about the sound of the trains asking if it annoys them and how they cope...and they suddenly realised they hadn't been aware of the trains for some time...and there was a therapist that moved house and it turned out that one of his neighbours has a dog that wouldn't stop barking...and he really didn't want to have to make a complaint but he found the constant barking day and night was such a pain...then one day he realised the barking had stopped...a few days after realising the barking had stopped he was out walking when he bumped into a neighbour and he spoke about how wonderful it was that the dog owner had either got rid of the dog or trained the dog to be pleasant...this neighbour asked him what he was talking about...the barking hadn't stopped or improved...and this therapist realised that what had happened was that he had habituated to the noise so that he filtered it out and stopped being aware of it...and a number of years ago a Scientist carried out an experiment. He was curious how factory worker that work in extremely noisy conditions seem to be able to talk normally to each other without shouting or raising their voices and they seem to hear each other perfectly fine...he asked a factory owner if he could spend sometime in the factory and even sleep in the factory with all that noise...the owner though this was an odd request but said he could...when the Scientist arrived he was aware of how noisy it was and how the noise hurt his ears and how he couldn't hear what anyone was saying...he spent the day in the factory and slept the night...when he woke up the next day he noticed something odd had happened...he was able to hear and understand what the factory workers were saying...his unconscious mind had managed to block out the old constant background noise...and there are many possibilities and experiences that your unconscious can use to help you to...*find your own answer*...and many children grow up and leave home...and often parents find their child breaking away from the family a painful experience...some parents *work through this pain quickly*...others take some time...yet parents learn that they need to *move on* and they...*let the pain go*...and you can *learn on an unconscious instinctive level what you need to do to*...*move on*...and...*create the changes you want*...and over the next few minutes you can take as long as you need to honestly and fully *create those changes*...and you can be curious about when you will be looking back and realising *the old problem has been gone for a while now*...(pause for about 3 minutes)...'

Performance enhancement – getting in the zone

'That's right...and you can really begin to think about those times you have naturally been *'in the zone'*...and so many people find they...*go in the zone*...without trying or without knowing...*it is going to happen*...and...*entering the zone*...is an unconscious process...and while you listen to me *your unconscious* can begin to access all the memories of times...*you naturally enter the zone*...and on a conscious level you don't have to be aware of those memories whilst the unconscious part of you begins to...*explore those memories*...and to...*look for how you naturally enter the zone*...and when...*you enter the zone*...you do so easily and effortlessly...when you put effort into...*entering the zone*...you increase the chances of not...*entering the zone*...and you can take this...*experience now*...to really delve into how...*you enter the zone*...and you know you can...*enter the zone*...with a trigger and you don't even have to know what that trigger is...or how the trigger works...*the trigger can happen automatically*...and you know you have experience of triggers happening automatically...like seeing a red light and putting on the brakes...or hearing your name and suddenly and instinctively paying your attention to where your name came from...and I wonder how *you will naturally and instinctively enter the zone*...when the time is right...and your unconscious...can *find a trigger*...many golf players...*develop a trigger*...of...*entering the zone*...when they take hold of a golf club...and many snooker players...*develop a trigger*...of...*entering the zone*...when they hold the snooker cue...and many F1 drivers...*set up a trigger*...and...*enter the zone*...when they hold the steering wheel...and I wonder how...*you will enter the zone*...and what your trigger will be...and a part of you can...*learn that instinctively*...and as...*you now learn that instinctively*...you can wonder what it will feel like...that feeling of...*focus*...that feeling that...*all of your attention narrows to the task at hand*...that all distractions disappear...that *time seems to slow down*...and you discover it's like *you're in your mind observing what your body is doing*...and Bruce Lee said 'I don't punch...it punches'...and you can now take the time to *create all the necessary changes* on an instinctive level...and you can take as long as is necessary for you over the next few minutes...to...make it possible to enter the zone state instinctively and effortlessly in those situations you want it to happen automatically in...(pause for about 3 minutes)...'

Performance enhancement – improving ability

'That's it...and as you continue to relax you can take some time to create an environment in your mind where you can...*learn what you need to know*...and there was someone that needed to learn for an exam in school and they took the time to...*create a place to learn in the mind*...and they went to that place and discovered a round table with people that were the leaders in their fields sat at the table and they sat at the table with these people and enjoyed conversations about their specialist subjects as they absorbed the learning...and another person wanted to learn to sing and dance at the same time...and they knew how to sing...and they knew how to dance...they just needed to learn to sing and dance at the same time...and they created a dance studio in their mind where a top instructor guided them through months of rehearsal...*learning...and practicing*...until they mastered singing and dancing at the same time...and you can *learn what will work best for you*...and now I'd like to have you just get the sense of a large high definition TV in your mind...and on that TV just get a sense of a slightly older version of yourself...and watch that slightly older version of yourself and you know what you want help with and what you want to improve and in a moment you can watch that slightly older version of *you perform perfectly*...everything they do will *be faultless*...and you can watch them even in challenging situations...*doing everything perfectly and effortlessly*...and I wonder how much you know about the research carried out into mental rehearsal that showed that people that used mental rehearsal without any actual practice performed better than those that spent the same length of time practicing...in mental rehearsal they practiced *performing perfectly* over and over again...whereas in real life practice people made mistakes and learnt how not to *perform correctly* as well as how to *perform correctly*...and it is important to *carry out real life practice* and you can discover how much improvement *you achieve each time you enter a trance*...and you can *drift into a trance before you fall asleep or as you wake up or while you stand in a queue or any other situation that is appropriate and safe to do so*...and each trance you have can enhance the effectiveness of this work towards achieving what you want...and I don't know whether it will be the spaces between my words or each out breath you take that helps you to...*improve performance*...and I wonder what will help that improvement continue...and you know it's interesting how as a child you thought differently and your thinking style allowed you to...*learn effortlessly*...and you made mistakes but they were part of the learning and they made *you learn even quicker*...and you learnt to master walking...learnt to master talking...learnt to master a whole range of complex processes tasks and activities...and you continue to do so...and in a moment you can take all the time you need to absorb integration and new learning and create and begin to *strengthen relevant neural pathways in the mind and body*...and your unconscious can draw up every single time you got a part of that correct and during this time can piece all of those parts together...in the same way someone can do this with golf and draw up that one time they made contact with the ball just right...that one time they had their feet just right...that one time they were thinking just right...the one time they got that swing just right...the one time they got the grip just right...and all these parts can be recalled unconsciously and linked together...and you can *honestly and fully carry this out*...*now*...that's right...(pause for about 3 minutes)...'

Learning to relax

'That's it...and you can take this opportunity to...*really absorb new ideas*...and you can instinctively *begin responding in a new way*...and as you absorb new ideas and instinctively begin to respond in a new way I wonder what other benefits you will notice...now...one way to...*really learn to relax*...is to begin to master 7-11 breathing...and 7-11 breathing is a way of breathing that creates real and rapid physiological changes...and each time you breathe in counting to 7 and out counting to 11...you begin to master the relaxation response...and whenever your outbreath is longer that your inbreath you trigger the relaxation response...and *feel good*...chemicals and hormones release through your body...and you can learn other ways to...*discover yourself instinctively relaxing deeply*...and one way you can...*discover yourself relaxing deeply*...is to master self-hypnosis or meditation...and you know the more you enter self-hypnosis or meditation the more relaxation becomes a part of who you are as a person...and as...*you master relaxation*...it begins to become more instinctive and...you find *relaxation happens all by itself*...another way to...*learn to relax*...is to allow your mind to wander and to think about pleasant experiences or even think about a pleasant time you have been hypnotised or drifted off into a trance...and the more the mind faces challenges the more *the body can relax*...and there are times when relaxation isn't an appropriate response and times when it is...and you know which times are which...and I wonder what else you know that you didn't know you know yet you knew you knew it without realising...as *you master the art of relaxing* when that is the ideal response to have...and you can learn here and now how to re-enter a hypnotic state and really...*learn how to enter a hypnotic state instinctively*...and feelings occur unconsciously first and I wonder whether your unconscious will decide to trigger relaxation when you begin to feel unconsciously stressed or whether your unconscious will wait until you begin to think you need to relax...and you can *discover that for yourself*...and each time you enter a trance you can *learn to relax* in your own unique way...and you can take some time now to make the necessary changes on an instinctive level to...*learn to relax...now*...(pause for about 3 minutes)...'

Removing a phobia with no known origin

'That's it...now what I would like you to do is to imagine that you are sitting in the World's most comfortable chair...a chair so comfortable you find all you can do in it is relax...and as you relax you can (face over to where you want them to imagine the TV and do this whenever you are talking about that TV so that you are being congruent with what you are saying) begin to notice a small old fashioned black and white TV...and at the moment the TV is off...and in a minute...while you continue to *relax deeper and deeper with each breath you take*...an old, old movie will appear on that black and white screen...and it will be an old movie of made up from all the memories that led to the development and maintenance of the phobia...and you don't have to know what any of those memories are...you can just relax and follow the process...and that collection of old memories will begin from a time before that old phobia response occurred...a time you were happy...calm and playful...all the way through that old memory to a time after that phobia had passed and you were calm and happy...and before that old distant movie plays I'd like you to get a sense of drifting out of your body...through space and time...drifting and relaxing to a position over beside that old TV...to a position where you can look over and see the you sat here but can't see what is on the TV...and when you have done that allow the head to nod...(wait for the head to nod)...that's right...and in a moment I want you to watch that you over there pressing fast forward on the TV remote when I say now...and watch that you sat there watching that old distant movie rapidly all the way through from that happy point at the beginning to that happy point at the end...and then when you have done that just allow the head to nod...and do that...now...(say these parts quickly)...that's it and really, really quickly fast forwarding that old movie all the way through to the end and allowing the head to nod...(watch for the head nod)...that's right...and now drift through space and time and float into the paused end of that old distant movie...drift into that happy paused end of that old distant movie...seeing what you saw...hearing what you heard and feeling what you felt...and once you have done that you can allow the head to nod...that's right...and in a moment when I say now I'd like you to rewind all the way back to that calm point at the beginning...and I'd like you to take no more than 2 seconds to do this...then allow the head to nod...and do that...now...that's it rewinding rapidly all the way back to the beginning...everything happening backwards the speech happening backwards...the action happening backwards and all the feelings happening in a new direction...that's it...(wait for head nod)...and now just get a sense of drifting through space and time back into that seat and in a moment when I say now I'd like you to watch that old, old distant movie all the way through to that calm end...then allow the head to nod...and take no more than a second to do that...and do that...now...fast forwarding that old, old movie all the way to the end taking no more than a second then allowing the head to nod...that's it...and now get a sense of drifting through space and time into the end of that old distant movie...seeing what you saw...hearing what you hear and feeling what you felt...and when I say now you can rewind all the way back to the beginning to that calm point at the beginning...and as you do everything can go in a new direction...people can speak in reverse...all the action can happen in reverse...and the feelings can go in a new direction...and you can do that taking no longer than the sound of a snap of the fingers...and then allow the head to nod...and you can do that...now...that's right...and now you can get a sense of drifting through space and time to that chair and as you relax comfortably in that chair when I say now you can watch that old distant movie through at a speed you are comfortable with...and when you have finished comfortably watching that old movie the head can nod...and you can do that now...that's right...and now you can get a sense of a you in the future on that screen...and the screen can become clearer and more desirable...and as it does you can watch that you in any future situations that would have led to you feeling uncomfortable in the past as that old you...and you can really begin to explore what is different when you respond in this new way...and you can *now take some time to relax and instinctively integrate and update your mind and body patterns*...(pause for 3 minutes)...'

Removing a phobia or PTSD with a known origin

(With phobias & PTSD it is often best to de-traumatise the worst or earliest memory first then the 2nd and 3rd worst memories if necessary. In many cases there may well only be a single memory that needs working with)

'That's it...now what I would like you to do is to imagine that you are sitting in the World's most comfortable chair...a chair so comfortable you find all you can do in it is relax...and as you relax you can (face over to where you want them to imagine the TV and do this whenever you are talking about that TV so that you are being congruent with what you are saying) begin to notice a small old fashioned black and white TV...and at the moment the TV is off...and in a minute...while you continue to relax deeper and deeper with each breath you take...an old, old movie will appear on that black and white screen...and it will be an old movie of that first or the earliest remembered phobia memory...and that old memory will begin from a time before that old phobia response occurred...a time you were happy and calm...all the way through that old memory to a time after that phobia had passed and you were calm and happy...and before that old distant movie plays I'd like you to get a sense of drifting out of your body...through space and time...drifting and relaxing to a position over beside that old TV...to a position where you can look over and see the you sat here but can't see what is on the TV...and when you have done that allow the head to nod...(wait for the head to nod)...that's right...and in a moment I want you to watch that you over there pressing fast forward on the TV remote when I say now...and watch that you sat there watching that old distant movie rapidly all the way through from that happy point at the beginning to that happy point at the end...and then when you have done that just allow the head to nod...and do that...now...(say these parts quickly)...that's it and really, really quickly fast forwarding that old movie all the way through to the end and allowing the head to nod...(watch for the head nod)...that's right...and now drift through space and time and float into the paused end of that old distant movie...drift into that happy paused end of that old distant movie...seeing what you saw...hearing what you heard and feeling what you felt...and once you have done that you can allow the head to nod...that's right...and in a moment when I say now I'd like you to rewind all the way back to that calm point at the beginning...and I'd like you to take no more than 2 seconds to do this...then allow the head to nod...and do that...now...that's it rewinding rapidly all the way back to the beginning...everything happening backwards the speech happening backwards...the action happening backwards and all the feelings happening in a new direction...that's it...(wait for head nod)...and now just get a sense of drifting through space and time back into that seat and in a moment when I say now I'd like you to watch that old, old distant movie all the way through to that calm end...then allow the head to nod...and take no more than a second to do that...and do that...now...fast forwarding that old, old movie all the way to the end taking no more than a second then allowing the head to nod...that's it...and now get a sense of drifting through space and time into the end of that old distant movie...seeing what you saw...hearing what you hear and feeling what you felt...and when I say now you can rewind all the way back to the beginning to that calm point at the beginning...and as you do everything can go in a new direction...people can speak in reverse...all the action can happen in reverse...and the feelings can go in a new direction...and you can do that taking no longer than the sound of a snap of the fingers...and then allow the head to nod...and you can do that...now...that's right...and now you can get a sense of drifting through space and time to that chair and as you relax comfortably in that chair when I say now you can watch that old distant movie through at a speed you are comfortable with...and when you have finished comfortably watching that old movie the head can nod...and you can do that now...that's right...and now you can get a sense of a you in the future on that screen...and the screen can become clearer and more desirable...and as it does you can watch that you in any future situations that would have led to you feeling uncomfortable in the past as that old you...and you can really begin to explore what is different when you respond in this new way...and you can now take some time to relax and instinctively integrate and update your mind and body patterns...(pause for 3 minutes)...'

Obsessive Compulsive Disorder Relief

'That's it...and you can begin to get a sense of a TV in your mind...and on that TV in your mind you can begin to notice a slightly older version of yourself...and you know there is a difference between what is appropriate and what is inappropriate and a part of you knows that difference...and as you watch that you on that TV you can begin to notice as they drift back in their mind to the initial memory that led to that compulsive need...and as you watch that you recall that experience you can watch as they receive all the appropriate and necessary therapy to make them better...and you can watch as they really recall that experience...and sometimes compulsive behaviours begin with a single event other times they have a number of events and the experience that leads to the compulsive behaviour is an event of thinking in a different way to that final event...almost like the straw the broke the camel's back...and you can watch that slightly older version of yourself as they explain to that younger version of yourself what the future will hold if they respond in a specific way and what they need to do to prevent that response unfolding...and you can watch as they help that younger you to...*keep the problem in the past*...and you know sometimes people worry about something happening again and then develop a habit of prevention...and this habit of prevention can be set at too sensitive...and while you watch that you being treated you can begin to unconsciously process what is an appropriate sensitivity that will let you know any risk is reduced to an acceptable level...and some people get really frustrated with their car alarm because it is so sensitive a leaf landing on the car triggers the alarm...and the owner has to keep going back to the car to reset the alarm...and it can be very frustrating and time consuming...and even when they don't hear the alarm going they are always on edge thinking at any time now the alarm could go again...and they feel so relieved when they have the alarm fixed or set to a lower sensitivity...and some children learn that plants need water and sunlight to live and grow...and because plants often start growing slowly some children get frustrated and think they must be doing something wrong because their plant isn't yet grown...and they will give the plant more and more water and the plant can't handle all the water and dies...other children think the plant needs more sunlight and because the plant hasn't grown enough yet they will give it more light...and then in the strong overwhelming sun the plant withers and dies...yet the child that let the plant do what the plant's got to do...and they *just relax* and have patience...and each day they check the soil is moist and only add some water if it is too dry...and they allow the plant light but not too much...and if they see signs the plant is getting too much sun they move the plant out of the sun a bit more...and they respond to the plant's needs...these children grow successful plants...and they find it more effortless because they haven't spent all their time worrying...and you can notice how that future you has been able to help that you from the past...and how they now begin to imagine future situations where that old behaviour would have happened and notice as it doesn't and notice how calm that future you looks as they watch a future you in those different situations...and at times that future you can watch some of the more challenging situations as they rapidly fast forward and rewind while they watch until they notice all the old emotion drains out of those situations completely...then they can watch the situation through at a normal speed *in a new way*...and in a minute you can begin to imagine what it would be like to *now be in that future you* watching those future situations...and *you have all the resources inside yourself* and you can instinctively discover how to...*respond differently*...and what it feels like to know...*you respond differently*...and many people...*develop a new therapeutic compulsive habit*...they find that they...*feel compelled to carry out this new therapeutic habit*...and...*discover it keeps the old habit away*...and they begin to...*become compelled to enjoy life and help those around you to enjoy life*...and they often begin to notice that none of what they were concerned about actually happened...they...*become comfortable with the low level of risk*...and often report that they...*feel like a normal person again*...and that they...*gain more control in your life by stopping the old behaviour*...and often people say they behaved in a compulsive way to achieve control yet on looking back discovered that the old OCD was controlling them and they just hadn't stood up against it...and in school there was a child that was being bullied and he would walk to school late and would take different routes and would keep away from people all just to stop himself getting bullied...and the more he tried to stop it by behaving in this way the more he stood out as being different...and the more he stood out as being different the more he got picked on and it was a downwards spiral and the problem continued to get worse...the more he behaved differently to try to prevent the bullying the more the bullying occurred...and his life began to revolve around avoiding being bullied...and he would walk many extra miles a week and he wouldn't socialise...and he wouldn't pursue any of his hobbies and interests...and all of his free time was spent worrying about being bullied and how he could prevent it...and he just couldn't see the wood for the trees...he kept following the same over the top solution rather than *finding a new solution*...one day this child spoke to an adult about his situation because he was now beginning to get a bit older and was realising that his behaviour is controlling him and so he now feels like he can't win...if he doesn't behave that way he gets bullied and if he does he is always anxious and frustrated and unable to lead a normal life...the adult gave him some good advice...the next day the child went to school much happier and was no longer a victim he returned to being himself...he put others first which led to some people taking advantage but he learnt from each experience...and he followed his interests...and he responded differently to the old bullies...he didn't become aggressive or appear anxious...he became indifferent to them...and after they stopped getting the response they wanted they got bored of trying to bully him and left him alone...he found a whole range of new positive changes that stayed with him as he grew up...and you can *get a sense of being that future you*...feeling different and *responding in a new way to situations*...and you can take some time now for a part of you to begin to reprogram the changes into the mind and body on an instinctive level...and maintain them with each breath you take...(pause for about 3 minutes)...'

Boosting confidence

'That's it...now you can begin to get a sense of a TV in your mind...and on that TV screen you can see yourself in situations that you feel confident in...and I don't know whether it will be six or seven or eight or more situations that go into creating that movie of confidence on that screen and each one can be like a separate scene will all of them playing one after the other and then cycling round so that at the end of the last scene you see the first scene begin again...and as you watch that a feeling of confidence can increase throughout those situations...and as you watch those situations you can begin to feel drawn into them so that you can see what you saw...feel what you felt and hear what you heard...and you can be curious about how you know you are confident as that feeling of confidence continues to cycle round and round with all the memories...and you can begin to add in brief clips of situations you used to not be confident in so that the feeling of confidence continues through all the memories and through these brief clips...and as the memories continue to cycle round the memories can begin to become brief clips while the brief clips can begin to extend whilst holding on to that confident feeling...and as you begin to continue to increase that confidence so the thumb and first finger of your right hand can begin to drift together and when they touch they can link the feeling of confidence with that action of having the thumb and first finger touching...and I don't know whether *it will happen automatically* in situations you want more confidence or whether you will consciously choose to bring that thumb and forefinger together...and as that confidence increases towards a peak the thumb and forefinger can drift apart again...and you can be curious about how *you will continue to improve your confidence* in the future...and in the future you can stand up and close your eyes as you imagine a circle on the floor in front of you...and you can imagine either yourself with that confidence or someone else that you know has that confidence standing in that circle...and you can watch them and pay attention to how you know they are confident...and turn up the volume to their internal dialog to hear how they talk to themselves that helps them to be confident...and then you can step forwards into that person so that you become that person...and you can then experience...*being that person*...and...*make that confidence a part of you*...and then in those future situations you can open your eyes and continue to hold that confidence as a part of you or find that the *confidence is fully programmed into the mind to suddenly occur in the future when it is needed*...and you know that a little bit of nervousness helps to improve performance and ability...and you can now just take a few moments to imagine a line of events in front of you from your future with the next situation that you would have lacked confidence in at the end of that line of events...and you know true confidence comes from being confident in your abilities...many people make the mistake of lacking confidence because they can't do something...or because they aren't as good as they think they should be...you can only be confident that you will do your best...and you can imagine that line of events...and you can begin to move towards that situation and as you approach you can watch it playing out how you want it to go...and as you approach I wonder what the natural trigger will be that turns on that confidence instinctively and automatically just as you enter that situation and others like it in the future...and you can get a sense of entering that situation and notice what it feels like to...*be confident*...and you can now become more absorbed in this learning experience of *installing that confidence in your future*...and you can take as long as you need while I quieten down in the background for a moment...(pause for about 3 minutes)...'

Enhancing motivation

'That's right...and many people think that they lack motivation without realising that they wake up in the morning...they get out of bed...they eat food...and they do many other behaviours that take motivation...and motivation consists of towards and away from motivation...there are things you are motivated towards...goals that you want...and there are things you are motivated away from...like not wanting to get in trouble...and to...*become motivated with the most efficiency*...you can have an element of towards motivation and an element of away from motivation...and you can think of something you feel motivated to do and as you...*think about that*...you can begin to get a sense of where that image is in your mind...is it in front of you...off to the left...off to the right...above you...below you...and you can think about how big that image is and whether you see yourself in the image or are viewing the image as if you are looking through your own eyes...and now you can get a sense of your mind going blank...and once your mind has gone blank you can get a sense of what it is you haven't been feeling motivated to do...and you can get a sense of where that is in your mind...whether it is above you...below you…behind you...in front of you...to the left...or to the right...and you can get a sense of whether you can see yourself in that image or whether you are viewing that image as if you are looking through your own eyes...and you can now begin to move that picture to where the motivated picture was and your unconscious mind can make the adjustments that programme that in your mind in the same way that you have the other motivational memories in your mind...and you can *fix that in place* as if each breath works like adhesive to stick that new way of representing those behaviours in place...and you can...*discover something new*...and...*respond in a new way*...and your unconscious can *make changes easily and effortlessly*...and you can begin to think about how things will be different for you when you find yourself motivated to do things that previously you felt less motivated to do...and as you listen to me you can respond more fully and when your mind wanders you can update those neural networks in the brain...to *respond in a new and improved way*...and I remember a story about a frog that was stuck in a rut on a muddy road...another frog hopped along and heard the frog asking for help...when he found the frog he tried to reach down to get him out...he couldn't reach so he asked the frog if he could jump out...the frog said it was too hard...the frog in the rut asked the other frog could you jump down let me stand on your back then I can get out...the other frog said then he would be stuck in the rut...and he didn't want that...the frog decided he would just have to leave that frog...and he continued to hop along the road...after a while the frog that was stuck in the rut hopped past him and raced off...the frog chased after him and asked I thought you were stuck in the rut...how did you get out...the frog replied...there's a truck coming...and you can *discover your inner motivation*...in a way that is unique to you...and you can take all the time you need to allow yourself to honestly and fully update your neurology...while I take a few moments to quieten down in the background...(pause for about 3 minutes)...'

Improving self-image

'That's it...and as you continue to *relax deeper and deeper* into that instinctively responsive state of mind you can begin to wonder how you will develop the ability to...improve your self-image...and you can begin to...*make changes on an unconscious level*...and you don't have to know how...*these changes will occur*...and you can imagine yourself standing in front of a window...and you can be looking in through that window and see someone that you know loves you...and you might not even recognise that person all you know is that they love you...and you can be curious about how you know that that person loves you...what do they do to demonstrate they love you...how do they behave...what do they think...how do they talk to you...and as you show a curiosity about this you can find yourself beginning to wonder what it would be like to...become that person...and you can now...*begin to get a feeling of that*...and as you begin to see the world through that person's eyes you can start to catch a glimpse of yourself and begin to see that you over there through this you's eyes...seeing what they see that makes them love you so much...and many people struggle to think of positive things about themselves when asked because all of those positive things are buried in who we are as people and for some people may not be immediately consciously available...and yet these people have friends and family that they know care about them and like them...and these people wouldn't be friends if they didn't like you...and you can think about what they would say they like about you...you know often others see the parts of us we don't pay attention to...and you can think about what others would say they like about you...and while a part of you thinks about what others like about you...you can understand and integrate what I am saying on an unconscious instinctive level...and a sculptor will take a large stone and see the beauty inside that stone...and that sculptor will keep that beauty in mind and will start chipping away...and they will sweat and work hard and struggle and at times they will want to quit...yet they will continue to...*chip away at it*...day and night...until they have removed all the unwanted stone...and they are so proud of what they have achieved and they take that work of art to a gallery to be displayed for everyone to...*admire the beauty*...and people...*admire the beauty*...and all the hard work that went in to creating that work of art...and *you really can*...now...*create your own personal work of art from the inside out*...in your own unique way...and looking at a garden you can think that the lawn looks beautiful and that all the grass looks neat and even and all the grass looks the same...yet when you get close to that grass and...*really pay attention*...you notice that each blade of grass is different...yet all of the blades together share a similarity...and all people on the planet share one similarity and that similarity is that we are all unique and different and beautiful in our own way...and you can begin to absorb the meaning in my words from the inside out...and you can take as long as you need now...to integrate all the learning and update psychological patterns deep within your neurology...and as you do I will quieten down into the background...and while I quieten down into the background a part of you can be wondering how all the changes will take effect and whether they will all take effect at the same time or whether some parts of the changes will occur faster than others...and you can wonder how that will happen for you and whether it will be different or the same as it is when it happens for others...(pause for about 3 minutes)...'

Setting and achieving goals

'That's it...and as your mind continues to wander you can integrate everything I say fully and honestly or discover in the future that everything I said became an integral part of who you become, and you can fall asleep while you integrate what I say or you can stay awake as you listen and absorb what I am saying to you...(pause for about 15 seconds)...Now talking to you in the future as someone that has achieved what you wanted you can begin to recap to yourself what your life is now like and as you do that I will quieten down in the background...(pause for about 2 minutes)...and I understand that you got where you are today by putting in time and effort and dedication and commitment...and that it all started when you set your mind to it back in (add the month/year) and that you then began to...*make all of those changes*...and continued those changes right through to the when you achieved success...and I'm sure you had a few hurdles and I'm curious how you overcame them...and now you are here in the future...aware of the events that have passed...knowing that you can drift back to the present, to the place on that journey you know you are currently at being aware of all that has passed and aware there is more yet to come, having knowledge that the present is a gift and that the future holds many more presents for you to discover...and you can relax in the knowledge that each time you drift off into a daydream or find your mind wandering...you are taking another step closer to making your dreams reality and as you continue to take a little longer just to finish integrating all this new learning into your neurology and into the instinctive part of you I will quieten down again into the background...and you can take as long as you need over the next few minutes to fully and honestly complete that...(pause for about 3 minutes)...'

Metaphorical 'life changer'

'That's it...and as you continue to *relax deeper and deeper* you can listen to me in the background...and as you listen to me in the background you can understand what I'm saying on an unconscious level...many years ago there was a prince that lived in a castle...and one day he was gazing out of one of the castle windows looking out over the land that his family ruled...and from his perspective he could see people struggling...he could see people suffering...he could see people starving...while he was in this castle getting whatever he wanted with the ring of a bell...the prince ask his father why things should be this way...and was told the world is the way the world is...there's nothing you can do to change it...the prince wanted to know why not...and asked what about all of those poor people...and we are so rich...can't we help them out...his father didn't answer...the prince thought about his life...he had everything money can buy yet he felt unhappy and felt like something was missing from his life...while many of the people in the town around the castle appeared to struggle and suffer yet they would often be smiling and looking happy...that night when everyone went to bed the prince snuck out of the castle in disguise...he had to *discover the answer*...how can people be so happy with such miserable lives...he walked through the town and into the nearby forest...the forest was dark and eerie and the night was still...there was shards of moonlight glistening down through the leaves of the trees onto the forest floor shimmering a path for the prince to follow...the prince followed this unknown path to discovery and wonder and was curious where it would lead him...as he continued to walk the prince could hear noises of animals and birds moving in the dark...he could feel his heart beating loudly as he continued into the unknown...after a while the prince found a clearing and saw a frog sitting by a pond...as he approached the frog it started to talk to him...*'you are on a journey of discovery and wonder and you can wonder what you will discover on that journey'*...the prince began to wonder as he continued to wander through the forest...he thought he could hear something behind him moving in the dark...following him...he could see the foot of a mountain and wondered how he would reach the cave...then out of the dark some tribes people appeared...the prince couldn't understand what they were saying but could tell they were friendly...they began to cut down some trees and build a ladder...the prince decided to help them...after some time they managed to complete the ladder and the prince climbed up to the cave...in the cave was a flickering fire...he sat down by the fire watching the flickering flame and noticing how the light from the fire was dancing on the walls...as he continued to watch the flickering flame he began to see things and his mind began to wonder...then everything went dark for a moment before he found himself resting beneath a tree...and as he rested beneath that tree he began to *discover the true meaning of happiness and life*...and he knew that when you get hot you enter a pool to cool down...and after a while you need to warm up so you go back out in the sun...he then got an image of the Yin Yang symbol...then a coin flipping in the air...then he saw a person struggling and while the person was struggling he saw the person step outside themselves and the part of them with the problem began to change how they were dealing with the situation...then the prince saw people socialising and talking with each other...different images and ideas drifted in and out of the princes mind...and with each image the prince was learning something new about himself that was beginning to change his life in ways he didn't yet know...and the prince continued to notice what seemed like random images until they disappeared when his unconscious knew it found the answers...(pause for about a minute)...and when the time was right he drifted back to that cave before climbing down the ladder and enjoying an adventure of learning and discovery as he headed back to the castle where he knew how the next day he was going to change his world forever...and improve the lives of those in the town...(pause for about 3 minutes)...'

Overcoming insomnia

'That's it...and you can be curious which night will be the first night...*you sleep the whole night through*...and I wonder how...*you will achieve that*...many people have found that the sleep problems they had were due to worrying too much during the day and going over things in their mind when they should have been *falling asleep*...and they found many ideas worked for them...and I'm curious which ideas will work for you...one person decided that if they weren't asleep within 20 minutes of going to bed they would get up and tidy their home...and within a week the home was tidy and they were sleeping through the night...another person decided to listen to meditation or hypnosis tracks as they were falling asleep to have something to follow that stopped their mind wandering and I worrying...and I don't know what will work for you...and I wonder whether...*you will sleep through the night*...tonight...or whether...*you will sleep through the night*...tomorrow night...or whether the first night...*you sleep through the night*...will be in a weeks' time or two weeks' time or a months' time or some time sooner...many people discover the...*improved night's sleep*...develops as any day time problems or issues become resolved...and you can resolve any issues or problems in your own way...and when...*your sleep improves*...I wonder how life will be different...and a part of you can take some time now to get a sense for what that life is going to be like...who will notice that...*your sleep has improved*...what will they notice...how will it make...*you feel different*...what other life changes will occur...and how will the change be maintained...and a part of you can drift off into the future looking back and reviewing how you achieved...*sleeping better each night*...and looking at those times when you had the odd brief blip and how you overcame those few odd blips...and when your head relaxes on the pillow I wonder how quickly...*you will fall asleep*...and some people find they fall asleep as their head hits the pillow...others find they fall asleep just after their head hits the pillow...and some people find they fall asleep after they have taken a few breaths...and I wonder what will work for you...and you can now take as long as you need to reprogram your neurology...while I quieten down in the background...(pause for about 3 minutes)...'

Overcoming impotence

'That's it...now a part of you can...*begin to make a difference*...and you have had experience of getting embarrassed and having your face go red and you have had experience whether you realised or not of blushing during intimate situations...and a part of you can learn how to stop blushing on your face when in those situations and can teach you how to send the blushing south...*well away from your face*...and that intense blushing can find a way of expressing itself and being noticed and *standing out*...because blushing is a signal that is trying to make you behave in a specific way...blushing has a message it wants you to be aware of...and you know some problems have an emotional component others have a physical component and others have both...and you can put old emotions in the past...you may as well...*forget about them*...and allow them to be totally attached to the memory they occurred in back in the past...almost like turning over the pages of a book...you know there is writing on the previous pages but can you remember the 192nd word on the page 10 pages back...and although you know the context of the story and could give an overview of the story you can also forget lots of information that isn't helpful to remember now...and as you take my words and allow them to stimulate your thinking and your neurology they can *thrust* new ideas and understandings into your mind and body...and as those ideas *grow and develop* so you can *learn unconsciously*...and you know it is easier to forget you had a problem than you realise...and you can just get absorbed and carried away in an experience...and it can be a pleasure to experience the intense feeling of a new discovery *that stands out noticeably*...and you have experience of your head standing upright...and you do that effortlessly...and when you get tired your head flops and falls...yet when you need to...*keep it up...it stays up*...and many people go to watch films in the cinema and eat popcorn...and they put their hand in the popcorn and then lift it up to their mouth...and then they lower their hand down...and they repeat this...until they get excited in a moment and the hand that is rising to the mouth suddenly pauses in the air and *stays stuck their until the excitement is over*...and in a moment I'm not going to talk to you...*I'm going to talk to your penis*...and when I do *your penis can listen*...and it can *feel a compulsion to respond fully*...and after you leave here today your *penis can remain in a trance and it will always remain in a trance responding to what I say*...as I talk to your penis now...and you know how to ignore (enter clients name) and do what...*you want to have fun*...and you can *be like a rebellious teenager*...you can do things for fun...you can make yourself known a few times every day...so that you don't feel ignored...you can begin to crave attention and want to *be noticed*...and you can *feel compelled to come out to play* whether (enter clients name) wants you to or not...and you can always *enjoy play time* and you can crave play and *take every opportunity to come out to play when you know it is right a proper and a decent time to do so*...and sometimes you can be cheeky come out at random times when you want attention...and I wonder how you will discover you keep intensely blushing and when you do the more you try to stop blushing the more intense the blushing becomes and you can work independently to what (enter clients name) thinks and just respond independently to situations...and you know which situations you can come out and play in...(pause for about 1 minute)...that's it and now I will talk back to you...and you can take as long as you need now on an unconscious level to integrate all the necessary changes to create the future you desire...and you can take as long as you need over the next few minutes while I quieten down in the background... (pause for about 3 minutes)...'

Overcoming enuresis (bed wetting)

'That's it and as you listen to this your mind can sort out that old problem in its own way...and you can begin to think about how life will be different once you have overcome that old problem...and I don't know when you will have your first totally dry night and whether you will have four or five nights where you stay dry most of the night before that first full dry night...and I don't know when you will first have two dry nights one after the other and whether this will be followed by a few wet nights before having another dry night or whether it will be followed by a few more dry nights...and I don't know when you will have a whole week of dry nights...and whether these dry nights start next week or the week after or maybe they start in a months' time...or sooner...and you know when a child learns to write they hold a pencil in one way and as they begin to...get used to controlling those muscles...these get better at writing and as a child learns to colour in pictures they start by going over the lines and as they...practice more and more...they get gradually get more and more of the colouring done inside the lines...and children learning musical instruments make many mistakes and gradually they learn how to perform all the movements correctly and link all the actions and...learn to control the muscles...and they improve and play better and better...and children that...enjoy learning...a sport have to learn how to make their body respond correctly instinctively...if someone learns football they have to master muscle control in the legs to make sure their foot is in the right place at the right time to kick the ball in the right way and with the right strength and in the right place...and someone learning baseball has to master muscle control in the hands and arms making sure they know where to move their hands to and when to close their fingers around a ball and how to get the strength of throws and direction correct...and you know how you would respond if you were using the toilet and someone walked in...*you stop peeing*...and *this happens instinctively and automatically*...and you can now take as long as you need to integrate and learn from all that I have been saying...and I will quieten down in the background for a few minutes to allow that integration to honestly and fully take place...(pause for about 3 minutes)...'

Life reprocessing technique

'That's it...and as you listen to the sound of my voice you can begin to explore new possibilities and as you explore those new possibilities you can discover changes occurring on an unconscious level...and as you listen to the sound of my voice you can begin to float and drift up out of yourself so that you are looking down on the that memory you were currently developing and as you look over from that position you can notice that all your memories are lined out before you stretching way back into the past...and you can begin to float back over all of those memories right back to the beginning...and in a few moments you can begin to work through every single memory of your life...and as you enter each memory you can save all the learning from that memory whilst healing any memories that need resolution...and you don't have to know how you do this...and as...*you work through each memory*...they can begin to *fill with a healing light*...and as you enter each problem memory you can notice that some memories have a pleasure or anger aspect that builds up in the memory before being cut short with another emotion that is anxiety based like embarrassment or fear or helplessness...that you want to get away from...and other problem memories will have anger or pleasure that was a problem...and other memories will be anxiety based...and you have all the skills...knowledge...and resources to resolve any issues and...*heal every memory*...and you can take plenty of time now to make each memory a present and work through all those presents until you arrive back at this present here and now...and you can take as long as you need to honestly and fully...*do that now*...while I quieten down in the background…and you can be fully done when I continue talking… (pause for about 3 minutes)…'

Comfortable birth

'That's it…and as you continue to relax into this state you can begin to explore the potential of your mind and body…and you know in many cultures it is the man that goes through the discomfort of labour while the woman just…*enjoy the sensation of birth*…and many people wonder how this can be…and in many other cultures…*birth is a comfortable experience*…the mother only has the expectations of feeling the experience and enjoying the experience yet is not accustomed to feeling any discomfort…and I don't know what changes or alterations your mind and body will make to allow you to…*enjoy the birth*…and many parents that…*experience a comfortable birth*…discover that they were paying all of their attention to the sensations of the experience and the pleasure and excitement they were feeling as they were giving birth…and you know it is possible to sleep from the neck down whilst remaining awake from the head up or go into a trance as a body whilst you remain alert as a mind…as you experience the birthing process…and you can integrate and absorb all that I say here to ensure *you have a comfortable birth*…and you can…*feel so comfortable during the birth*…and…*enter a healthy state of mind*…and over the next few minutes you can take as long as you need now to really honestly and fully make all the necessary changes at an unconscious and instinctive level…(pause for about 3 minutes)…'

Overcoming fibromyalgia

'That's right…and as you continue to breathe in that way a part of you can begin to go back and review the past…and that part of you can enter every single memory and work through each memory…and as that part of you searches back through memories it can begin to *heal any memories* that…until now…have negatively impacted on the future…and some memories will contain anxiety feelings or difficult stress and others will at first appear to contain an emotion like guilt or embarrassment or fear or anxiety yet on closer inspection they will also contain anger or pleasure that until now you had forgotten all about…and this anger or pleasure feelings will have been suppressed or unable to be expressed at the time…and I knew someone that had a memory like this…and they remembered how embarrassed they were when they got caught dressing up and completely forgot the pleasure they felt before they were caught…and someone else I knew got bullied and they remembered being scared and feeling helpless but had forgotten how angry they felt and at the time they couldn't act on this anger…and on recalling these hidden or forgotten feelings both these people *recovered*…before the end of the session…and you will retain the learning made in all the memories that part of you works through…and as that part of you goes back through those memories…you can begin to discover a greater sense of comfort…and you can begin to feel like a weight is being lifted off you…and as that weight gets lifted off you so you can experience a lifting off of you of another kind…and while that part of you carries on working through those old memories another part of you can take as long as it takes between now and when you exit this trance to go through the whole body…through every muscle…every bone…every organ…and every cell…*applying a healing gel*…and this healing gel works by healing and preventing signals of discomfort…while at the same time having the properties to allow new signals unrelated to the old problem through to conscious awareness to allow you to gather further information…and a third part of you can begin to drift off…into the future and explore what the future holds…and explore how these changes you are making…now…impact on that future you…and how…*your life is different with these changes*…fixed in place…and you can take as long as you need to *complete all these three tasks on an unconscious level* before integrating fully back together by the time I next speak…(pause for about 3 minutes)…'

Acne removal

'That's it…and as you relax in that way your unconscious can begin to work in a new way…and you can begin to alter how oily the skin is and you can begin to make all the necessary changes to...*make spots vanish*...and you can make all the necessary changes to...*keep the skin clear*...and occasionally a rogue spot may appear and I wonder if you will have a chance to notice it before it fades away again...and you know when washing up is left to pile up it can seem like such a chore...and so much hard work...and all the food is dried on...and to make the job easier you scrape off all the loose bits of food and then put the washing up in warm soapy water to soak...when you come back after a few hours and drain the water and clean the plates with fresh soapy water the plates clean easily...and you can learn on an unconscious level what you've got to do...and I knew of this boy that wanted to...*get rid of acne*...and he had tried dozens of different medication...and none of it worked...and he was set the task of going on holiday to a log cabin in the alps with his mum and the log cabin was bare...there was minimal furniture...no pictures or mirrors on the walls...no TV...and no mobile phone signal...the boy and his mum were on holiday for a fortnight and no other intervention or therapy was carried out yet the boy came back with *clear skin*...and the boy didn't even realise that being in that log cabin was going to change his behaviour without him noticing and that this change of behaviour was going to lead to his...*face becoming clear*...and your unconscious can *make all the necessary changes now* and can understand this story on a deeper level and it can do that...Now...(pause in the background for about 3 minutes)...'

Overcoming anger

'That's it...and as you continue to relax in that way I will begin to talk to you and what I say to you, you can understand on a deeper than conscious level...and you can really begin to explore some issues...and you can do this on an unconscious level...and as you work through those issues I will talk to you...and you know when a car alarm is set too sensitively it goes off all of the time...and a leaf lands on the car and the alarm goes off...and someone walks past the car...and the alarm goes off...and you know if you don't set the alarm the car may be at risk...and the alarm can be set to an appropriate level and it goes off only when it is genuinely at risk...and there was a story I heard many years ago about a woman that wanted to help her husband to control his anger went to the village medicine man…and the medicine man told knows a way to help…but before receiving the answer she must three whiskers from a tiger…and on the first day she tried to appro the tiger but the tiger growled and lashed out at her…the second day she sat away from the tiger and didn't approach…on the third day she sat a bit closer and again spent the whole day just remaining still…after a week she had managed to get close enough to hear the tiger breathing…by the end of the ninth day she managed to get close enough to feel the tigers breath…when it approached the fourteenth day she was able to rest against the tigers belly as he slept…on the fifteenth day she gave the tiger food and did the same on the sixteenth and seventeenth days…by the eighteenth day she had managed to pick one whisker from the tiger…and the tiger flinched…on the nineteenth day she picked the second whisker and on the twentieth day she picked the third whisker…when she went back to the medicine man and told her what she had done and that she is ready for the answer now…the medicine man just told her…*you have found the answer…you no longer need my help*…and the woman went away confused and the next day her husband had changed and was…*now a much calmer person*…and you can learn from this and…*find the answer*…taking all the time you need over the next few minutes…Now…(pause for about 3 minutes)…'

Improving the nervous system

'That's right...and as you continue to listen to the sound of my voice...*your unconscious*...can begin to work on a deeper level...*your unconscious*...can begin to focus on the nervous system...and as your unconscious begins to get right down to a microscopic level and work on analysing and working through the nervous system it can begin to make any necessary repairs or overcome any blocks...and...*your unconscious...now*...can also begin to work through the settings in the various aspects of the nervous system to ensure that all responses from each part of the system works in a harmonious nature with each other part of the system...with each part of the system ensuring it can respond to stimulus and signals just the right amount...and each neuron in the mind and body and each nerve cell can

can be a time consuming task over the next few
)ugh every single nerve cell in the entire mind and
ying out an MOT or service check...and there are
:e to those neurons that outnumber the neurons
role and they can undo go a service check as
ns associated with sensory information
..uns associated with motor or movement based
₋ᴊɪɴg...and neurons throughout the central nervous system and any connections can all be checked to ensure they work correctly and any damage can be worked on and the pathways can be checked over to ensure clear free running and transmission of messages...and you know when you hack a path through a forest it always seems difficult the first time...after you have been through that path many times it is easy to follow and find...and you know flowing water always finds the easiest way through and water is a soft substance yet it can erode rock...and cut channels and pathways where previously there weren't any...and your unconscious can understand what I say in a new way...in a way the conscious mind just doesn't get...and you have a sympathetic nervous system that control the fight or flight response and a parasympathetic nervous system that controls the relaxation response and...your unconscious...can ensure these systems are in balance...in the same way that Goldilocks liked things just right...and you know more about your own nervous system on an unconscious level that either I do or you do consciously...and you can now take as long as it needs to take over the next few minutes to honestly and fully work through the nervous system to ensure it is working correctly throughout... (pause for about 3 minutes)...'

Improving the endocrine system

'That's it...and as you continue to relax in your own way while you listen to the sound of my voice your unconscious can begin to take a journey through your endocrine system...it can take a journey through the whole mind and body and through all related systems...organs and processes...it can work through ensuring that chemicals and hormones are able to be released in the correct quantities at the correct times...and your unconscious...can *learn from this journey* about how...you can help yourself in ways that are unconscious and unexpected...and we all need a recharge from time to time...we all need to make sure that things are working properly and that we can respond instinctively in the correct way at the correct time...and your unconscious now...can honestly and fully take time to make sure your mind and body are working in harmony with you...and you can be curious on one level if you will notice any of the positive benefits...and while *your unconscious*...works on all that it needs to complete my voice will drift off into the background...(pause for about 3 minutes)...'

Improving the digestive system

'That's it...and while you listen to my voice and relax with each breath...your unconscious mind...can begin to explore and adventure through the digestive system...and it can start with the mind...and start with the thought processes and the neural pathways and neurons that control the digestive system...and your unconscious mind can work through this process healing where necessary as it goes...and that part of you can take as long as it needs to complete this fully...while I will continue to talk to you...(pause for about 1 minute)...that's it...and as that part of you continues on its healing journey it can begin to work through the whole digestive system...and I wonder what kind of a dream or mental adventure it will create for you as it carries out its work...some people find they start to imagine a journey in a car travelling along motorways and down smaller roads and windy roads...and your just enjoying the view and enjoying the journey...and other people imagine something completely different...as *your unconscious*...now...works fully and honestly outside of your conscious awareness...and that part of you can check everything necessary to ensure the experience is a smooth one and to ensure all related and relevant organs are also working fully and correctly and processing correctly...and you don't have to know how any of...*this is taking place unconsciously*...you can just wonder whether you will notice the improvements straightaway or not...and you can take as long as you need to do this fully and honestly over the next few minutes...(pause for about 3 minutes)...'

Improving the circulatory system

'That's right...and you can make changes without even realising...and with each breath you take I'd like to have you get a sense of breathing in a healing light...and I don't know what colour that light is...and as you breathe out you can get a sense of breathing out problems...blocks and impurities...and as...*you breathe in that healing light*...you can get a sense of it passing down into your lungs...and as it does it can begin to cleanse and heal...and that cleansing and healing can continue as that healing light continues on its journey...and that light can fill the lungs and permeate each and every cell...*healing*...and repairing...and that healing light can continue to follow the route into the blood and throughout the body...and be absorbed by the body...and it can take as long as it needs to fully and honestly work all the way round the body removing any blocks in the circulatory system and adding extra healing resources to anywhere they are needed...and you can wonder consciously whether you will notice the positive changes and whether they will be noticed quickly or over time or whether you will even attribute the positive changes to this work your unconscious is doing here for you now...and you can take as long as you need over the next few minutes to make all the necessary changes and in the same that a broken arm only needs to be put in plaster once to allow the healing to occur over time...so that healing light only needs to start the healing once for that healing to continue and complete over time...(pause for about 3 minutes)...'

Overcoming Irritable Bowel Syndrome (IBS)

'That's it...and now as you continue to relax I would like to talk to that part of you that has been responsible for developing that old IBS response in the past...and I would like to thank that part of you that created that in the past for taking it's time and energy to help to convey an important message...and you know now...*you can stop*...and think of a new more pleasant way of getting that message across...and I know you have all the skills and abilities to...*develop that new response*...in a creative manner...and I wonder what it will be...and many people remember the frustrating experience of trying to untie knots...and the harder you try to...*untie the knot*...the tighter the knot seemed to get...and many people have learnt that to...*untie the knot*...they need to...*relax and take things slow and easy*...and carefully they can...*untie that knot*...and sometimes trees grow large in a garden and need to be removed and you can think long and hard about how to remove that tree or you can worry about what will happen if you don't remove the tree or you can take action...hire someone in to help you remove the tree carefully branch by branch until it is level with the ground...and then carefully dig out the roots one at a time...before planting a wonderful healthy flowerbed...and your unconscious can understand things in a way the conscious mind over looks...and you can wonder whether you have had IBS for the last time...or whether you may have it once...twice or even four times more...before...*you never have it again*...and you know change can be so quick when...*change happens instinctively*...and you can get splashed by a car and you know to avoid puddles when you see a car coming for ever into the future...and that happens instantly without any thought...and *you can learn* and...*heal yourself*...that quickly...*here and now*...and over the next few minutes you can take as long as you need to *make all the necessary changes* to ensure that old response and old problem remains just as a distant memory...and I don't know whether you will have all the work completed before the end of the session or just after the end of the session or tomorrow or next week but all the necessary changes can be completed and in place over the next few minutes while my voice drifts off elsewhere allowing you to work fully and honestly on an instinctive level...(pause for about 3 minutes)...'

Boosting the immune system & fighting infections

'That's it...and as you listen to me a part of your mind can begin to access your immune system on a deep level...to...begin to help it to recognise and fight infections and illnesses better...and it can learn from messages coming from the external world and understand that not everything that is different or alien is bad...and some things are actually on your side and there to help you to...*fight the true invaders*...and the conscious mind can think about what it wants help to fight...and what it wants left alone to do its job...and *fighting infections efficiently and effectively*...doesn't mean that the immune system needs to become hyper-sensitive...it can become appropriately sensitive...in the same way that many unnecessary allergies are caused by a hyper-sensitivity...and *your unconscious*...can understand this...and a part of you can now travel through the body placing a coloured substance or marker on all those bits that need eliminating...and I don't know whether that part of you will place the markers on the relevant places from the feet up or from the head down...and as that part of you responds in its own unique way...you can wonder how quickly positive change will occur...and while...*change occurs on a deep instinctive level*...I will quieten down in the background...and you can take as long as you need to honestly and fully...*make all the necessary changes*...over the next few minutes...(pause for about 3 minutes)...'

Increase fertility

'That's right...and you can...*begin to learn instinctively*...how...*it is possible*...to...*increase fertility*...and you don't have to know consciously how...*you do that now*...because the necessary changes can all take place on an unconscious level...and I don't know which changes will be physical changes and which will be psychological changes...and whether consciously you will be aware of any of the changes or whether...*all the changes will happen outside of your awareness*...and a part of you can instinctively go through all of your life history and experiences resolving negative emotions and blocks that were relevant to maintaining that old fertility response...and you don't have to know consciously what is going on unconsciously...it can just take place instinctively in its own way...and you know it is interesting...the harder you try to do something the more difficult it becomes...and the more desperately you want something the more elusive it becomes...and the unconscious part of you can focus its energy on helping you to...*correct any physical or biological blocks to effective fertility*...and I don't know whether you will reach the peak of fertility in one weeks' time or in 6 days' time or in 9 days' time or sooner...and when you do you can feel uncontrollably compelled to appropriately act on this moment...and you know some seeds can fall in a desert and remain dormant for years...and any observers would assume that the dry and barren desert has no life...yet when even a single rain drop comes the desert begins to flourish with life...as the seed soaks up the drops triggering new growth...and even if the soil isn't ideal to grow in *that new life takes hold and grows*...and water from one place can develop into a rain shower elsewhere...and if there is significant energy exciting the rain drops it can turn into a torrential downpour or a beautiful and powerful thunderstorm...and cactus plants grow in the desert despite the lack of rain...and they can grow up big and strong...and these desert plants live in a stressful environment yet they still appear and grow and survive...and in locations with heavy rainfall and very little stress life flourishes and grows easily and effortlessly...and *your unconscious*...can *learn how to generate fertile soil* and keep the conditions just right...and you know...*life grows*...in places that seem inhospitable...and...*your unconscious*...*now*...can integrate all that is necessary...*your unconscious*...can integrate changes in the mind...in memories...in beliefs...in the way you respond to stress and in the body...in all parts of the reproductive system and the reproductive process...and in each relevant cell in the body...and as...*change happens*...taking as long as is necessary over the next few minutes...I will quieten down in the background...(pause for about 3 minutes)...'

Treating psoriasis

'That's it...and as you continue to relax I will talk to you...and as I talk to you, you can listen...and while you listen your unconscious can begin to make changes throughout the mind and body...and *your unconscious*...can make these changes by checking to see what changes need to occur...and your unconscious...can start deep within the cells of the body and gradually take it's time to work from deep within the cells to how the body fights infection and whether this is happening appropriately all the way up to how you respond to stressful situations...and while your unconscious...does that I will talk to you in the background about the different causes of psoriasis...and some psoriasis has been known to have a genetic cause...and this genetic cause doesn't always lead to psoriasis...sometimes it can be lie dormant until a stressful event brings it to life...and in the same way that stress can bring it to life...relaxation can turn it off...some people have overactive white blood cells that normally fight infection but because they are too active the chemicals they release make the skin cells multiply too fast...and they can normalise themselves...normally skin cells go through their life cycle every 30 days...yet in psoriasis that is caused in this way they can go through their life cycle in 3 to 6 days...and they can...*relax*...and often the way the person responds to stress can start or maintain psoriasis...and I don't know whether you will respond differently to situations as the psoriasis disappears or whether you will think you are responding the same whilst the body responds differently as the psoriasis disappears...*and your unconscious now*...can take as long as it needs over the next few minutes to...make the change happen...and it can be like a snowball being rolled down a hill or a stone thrown in a pond...and I don't know whether...*the skin will clear*...during the day or during the night...and whether you will notice it in the morning or in the evening...when you wake up or when you go to bed...or whether someone else will be the first to notice...and whether it happens quickly over the next day or two or whether...*the change happens*...slowly over the next few days...and you can be curious to...*discover that for yourself*...(pause for about 3 minutes)...'

Self-awareness process

'That's it...and you can be curious about awareness...what does awareness really mean...what does awareness mean to you...what can you learn by...*enhancing awareness*...and you can begin to get a sense of drifting and floating up above yourself and see your life unfolding from the past to the present to the future...and you can observe how you respond to those around you...and how others respond to the way you respond to them...and sometimes we say and do things without really knowing the impact of our actions...and you can observe the impact of your actions...and you can get a sense of how you fit in with the world around you...how *you influence the world around you*...and you can discover more about yourself as a human being than you knew you knew...and you can be aware of every little action and interaction and the consequences of those actions both positive and negative...and you can really begin to know yourself on a deeper more fundamental level...and you can delve deeper in to what makes you you...what your beliefs are...what your traits are...what your skills and abilities are...what annoys or irritates you...and you can learn something about yourself and how you can apply your discoveries to *being the person you want to be in the future*...and my voice is just going to drift off in the background for a time and when it does you can begin a process of self-exploration and discovery...and I don't know whether this will happen quickly or slowly and whether you will have conscious awareness of this happening or just unconscious awareness...and the longer my voice remains distant the deeper and deeper into the experience of self-awareness and discovery you can go...and you can do that...Now...(pause for about 5 minutes)...'

Self-acceptance process

'That's it...and you can begin to explore your own natural ability to...*accept yourself*...and to...*accept yourself*...is to be comfortable with who you are as an individual...and you know every blade of grass in a meadow looks different on close inspection yet together a meadow can look beautiful...and you know the one thing all people have in common is that all people are unique...and I wonder what will be the trigger that makes self-acceptance possible...and you can step back and view your life...and as you view your life you can wonder how you got here...and you can look back over good decisions and bad decisions and wonder how they have shaped who you have become and who you will be developing into...and mistakes are unplanned lessons...and I wonder what you can learn from your mistakes...and how this learning can help to *enhance your own self-acceptance*...and as you review the past you can learn about the future...what will it mean to you to *have self-acceptance*...how about being accepted by others...how will...*things change*...and a part of you can instinctively explore these areas...and you can learn and you can understand and you can imagine a positive role model that accepts them self...and what is different about that person...what do they do...how do you know they accept them self...and you can be curious about what changes will occur that lead to...*you accept yourself*...and curious to discover how this affects who you are as a person and what will be different about the way you interact with others and develop over time...and I wonder if you can be curious about how people accept each other and put each other's faults aside...and how they manage this...*and your unconscious now*...can take as long as it needs over the next few minutes to make *all the necessary changes on an instinctive level* to...*discover how to accept yourself*...and find that *change happens*...as I quieten down in the background...(pause for about 3 minutes)...'

Enhance intuition

'That's right...and I wonder if you have ever had the experience of just knowing something but not knowing how you knew it...or just having a sense of something but not understanding where that sense came from...*and your unconscious understands*...and your unconscious understands how this happens...and have you ever looked back and thought to yourself that you saw a situation developing...*and your unconscious can learn from these types of experience*...and that instinctive part of you can now begin to draw up on all of these resources where you have...*experience a heightened state of intuition*...and you can begin to integrate this into who you are at an instinctive level...and there are times when you may not want to be so intuitive and times when you want to...*crank your intuitive abilities up to their full potential*...and to...*become instinctively intuitive*...you can begin to develop the ability to...*notice patterns*...and...*notice fine detail*...and you don't have to consciously know...*you notice these things*...it can happen unconsciously and instinctively all by itself...and you can take as long as you need over the next few minutes to really deeply integrate all that is necessary into creating the required changes to instinctively enhance your intuition and you can learn from past experiences where intuition was enhanced naturally and you just knew things...and you can be curious how much your intuition will increase and at what point it will become noticeable and whether anyone else will notice and if they do I wonder what they will notice...(pause for about 3 minutes)...'

Forgiveness process

'That's it...and as you continue to drift in this journey my voice can continue with you as a comforting part of your experience...and as you just go with this now...you can continue to enjoy this process of change...and just be aware of a bench just off in the distance...and just get a sense of walking towards that bench...and just allow yourself to sit down comfortably on that bench *deeper and deeper* into the experience as you sit down on that bench...and just look around and notice the view from that bench...notice what you can see...what you can hear...and notice what it feels like...sat on that bench there...and then just get a sense of drifting up out of your body...drifting up out of your body and floating over and settling down behind some bushes off to the side of that bench so that you are looking over at that you sat on that bench...and just notice how much more absorbing the experience can be...with that curiosity looking at that you wondering what is going to happen next...wondering who is going to come over and sit next to you...and as you look over at that you sat on that bench...just get a sense of somebody coming over...you don't have to know exactly who that somebody is...but somebody wanting forgiveness...somebody who is wanting to be forgiven...just get a sense of watching that you over there on that bench doing what they have to do to forgive that person who is sits down with them...and you don't have to just be aware of that you sat over there with that person they are about to forgive...and you can also be aware of something odd that happens...a light seems to come up off the shoulders up into the sky...just seems to *lift as that forgiveness is given*...and you know as the story goes that one of the criminals on the cross next to Jesus asked for forgiveness and Jesus gave that forgiveness and that man got the chance to go to heaven...and you know in all religions forgiveness is a huge part of that religion...to *have the strength to be able to forgive somebody for whatever they have done*...and just get a sense of watching that you over there *forgiving that person* that's sat down with them...I don't know if they are sat a few feet away or if they are sat right beside each other...just watch what happens as that...*you forgive that person*...watch the positive outcome of it...watch how you know the moment...*they are forgiven*...and then just get a sense of how you know that you there has just done the...*forgiving that person*...just notice how you know they are now happier as a person...just get a sense almost like you have got x-ray vision and you can see into them...and you can see that the way your genes are expressed are changing to calm relaxed settings...that not only has that you done something good for yourself but also done something good for future generations...and for those people that that you interacts with...that's it...and then as this you over here watching that you over there just get a sense of thinking wouldn't it be nice to know what it felt like for that you over there to do what they just did...and as you think that to yourself just get a sense of drifting into that you over there before that other person had come out and joined them...and perhaps there is butterflies in the stomach from anticipation because you know someone is about to come out and join you and you know you are about to forgive them...and your nervous about what it's going to feel like...how that pleasant feeling is going to happen when *you forgive that person*...what it's going to feel like to have that light lifting off your shoulders...what all the physical changes throughout your body will feel like...and changes in the mind...so just get a sense of that person coming out and sitting next to you and asking to be forgiven...and get a sense of forgiving that person and *letting go*...yet remembering you're not forgetting what they did...you are allowing them to own their own problem and live with what they did...and you can know that within yourself you are making the right decision leaving them with their guilt and leaving them holding full responsibility for their own actions...and after that light has lifted off your shoulders and that person has got up and walked off...just get a sense of leaving the bench and all that behind you as *you drift deeper into your mind to learn from this* experience preparing to exit trance with a pleasant change...(pause for about 1 minute)...'

Finding inner peace process

'That's it...and as you continue to relax in a moment I'm going to quieten down in the background...and when I do you can begin to *find true and honest inner peace*...and when *you find that inner peace* you can be curious how it will translate into your everyday life...and it can have a deep influence throughout your mind and body...and the interesting thing about silence is it's healing nature...not just healing physically but also its ability to heal you mentally and spiritually...and as I go quiet it can seem like you are walking through a door into a room of nothingness...into a room where the walls absorb sound preventing any echoes...and in that place of peace it can seem like you are a mind floating in space without a body...and as you drift and dream and float...with each breath...you can be aware of sounds without hearing them and be here as a body whilst you remain there as a mind...and you know when you sleep and dream...you completely and comfortably separate from physical reality for a while...and the more silence you experience the more inner peace you can discover...now...(pause for about 5 or more minutes)...'

Developing assertiveness

'That's it...and I'm wondering whether you would be able to say no to me...and how you know when someone is assertive...and I wonder how you know the difference between assertiveness and aggressiveness...and I would like you to get a sense of a person in your mind that is assertive...a person you know shows no signs of aggression at all...someone that is kind and caring and supportive...someone that thinks about others and how they can be helpful to others...and as you think about them just get a sense of what they look like...get a sense of how you know they are assertive...perhaps you can be watching them in action...what is it they do that lets people know they mean what they say...how do they behave...what body language do they use...what is their tone of voice...what words do they use...how do others respond to them...if you turned up the volume to that person's internal dialogue what do they say to themselves...really explore how that person does assertiveness...now watch that person being assertive as if they are standing in for you...watch them in future situations in your life and watch how those situations go differently to how you thought they would if it was you dealing with them...that's it...now take some time to start those future memories again...only this time step into that person so that you are seeing what they see...hearing what they hear...and feeling what they feel...and experience being them in the future...and notice what it feels like to experience things in that way with those responses...what it feels like to be assertive...and as you take time to experience a wide range of future situations you can spot a mirror and see that that person is actually you and has been the whole time...and that that you has learnt from that other person how to respond assertively...and before you integrate that ability to be comfortably assertive fully into your life I'd like you to take some time to look into the future at situations you don't want to be overly assertive...because there are situation where it is best to hold back or negotiate more or allow others to be assertive with you...and a whole range of different situations and circumstances...and you can take some time to explore those and how you would like to respond in those situations and then explore those situations where *you can comfortably and appropriately be assertive*...that's right...and now you can just take as long as you need to *make this assertiveness instinctive*...and you can also draw up all past successful experiences of being assertive and take as long as you need over the next few minutes to honestly and fully instinctively integrate this assertiveness...now...(pause for about 3 minutes)...'

Overcoming addiction

'That's it...and *you can overcome addiction*...in your own unique way...and in your own time...and you don't seem like the kind of person that likes to be bullied and pushed around...and you can *learn deeper ways of responding*...and a part of you can explore what had been keeping that addiction going...it can explore whether it is a pattern of behaviour where the cause is back in the past and no longer applies but has left behind a habit...or whether the addiction had been your own form of therapy where it was meeting a need or helping you in some way...and you know sometimes when a child gets a new toy they stop playing with their old toy...and they haven't lost the old toy...they still know where it is...they just don't want to play with it anymore...and I wonder what new healthy toy you will get...and all the changes that will happen will be in accord with who you are as a person and will have a positive influence on yourself and those around you...and you have experience of not having the old addiction...you have the experience of living life differently...and I wonder how...*that old behaviour will become redundant*...and that old addiction has been bossing you around...it has been manipulating you on a deep level...it has been controlling you and influencing your mind and decisions...and it is time to *take back control*...and you don't have to know how to do this consciously as the *changes take place on an instinctive level*...and the benefits can be achieved in new and exciting ways...and I'd like to thank that part of you that was kind enough to create that old addictive behaviour back when it was the only solution you thought was open to you...and now would like your help to develop this new and healthy response that will enhance your life and the lives of others around you in ways that may not immediately be obvious...and all the necessary changes can happen fully and honestly from breaking old unhelpful patterns and processes to getting needs met in healthier ways...to developing new behaviours and patterns and building on strengths and resolving issues and moving forwards...and your unconscious can do that in your own unique way...now...(pause for about 3 minutes)...'

Hay fever relief

'That's it...and now on an unconscious level you can begin to change the sensitivity settings to particles getting in the nose...and you can dull down the sensitivity...and your unconscious can begin to learn how to...*respond appropriately*...and your unconscious knows fully and honestly exactly what it needs to do to...*stop that old response*...and it can...*do that now*...in its own way...and it can take all the time it needs to...*make those changes*...and while...*your unconscious makes those changes*...you can begin to wonder how life will be different...what will life be like...how will...*you notice the difference*...who else will notice and what will they notice that lets them know...*you have stopped having that old hay fever problem*...and what other knock on changes will occur...what will you be able to do that you previously hadn't been able to do...and you can fully and deeply explore these ideas...and really honestly imagine them in the future and how life will be without that old response...what a difference it will make...and while a part of you does that another part of you can make all of those changes on an instinctive level...and you know a car alarm set too sensitively is really irritating and needs to be reset more appropriately...and as you make all the necessary changes to that instinctive and unconscious part of your mind I will quieten down in the background...and when I next talk to you, you will be just beginning to finish up making all of those necessary changes...(pause for about 3 minutes)...'

Enhance orgasms

'That's right...and I don't want you to experience too much pleasure just yet...really *save it for when the time is right*...and when sexual experiences begin I don't want you to experience too much pleasure too quickly...really hold off and wait until...*the time is right*...and it is important that you don't...*overdo taking on board my suggestions*...but take them on board just right...because it is all too easy to...*overreact or be taken by surprise*...and find you didn't wait...and you know that an intimate part of your body can blush strongly and uncontrollably when certain areas are touched...and any old blocks created by old experiences can become detached from those intense blushing moments...and those blushing moments can happen at a time you want to feel pleasure or excitement or love...and you can instinctively discover those areas that get touched to trigger that intense blushing...and I want you to really try hard to wait at least a week before letting that blushing occur...and when that intense blushing occurs I wonder what feelings you will discover that go with it...and the area that experiences that intense blushing can significantly increase in sensitivity...and the mind can *feel deeply connected to those experiences*...almost like...*you enter an orgasm trance*...and with each breath you take in that special way whilst you have sexual thoughts in mind deepens and focuses that trance...and the longer...you try really hard not to let it happen too quick...*the deeper into that trance you go*...and your unconscious can take this time now to make all the necessary changes honestly and fully on an instinctive level while I quieten down in the background...(pause for about 3 minutes)...'

Overcome premature ejaculation

'That's right...and it is possible to enjoy feeling things without acting on that feeling...and you know how to *hold back whilst having fun*...you know how to keep the punch line until the time is right...and I know you know how to hold off ejaculating by 1 second...and you know that would be easy and I know that would be easy...and I bet you could even hold off by 3 seconds or 5 seconds...and you have been giving this compliment and you can complement yourself on that...and you know some people struggle to ejaculate and they try and they try and they find it really difficult to do...*imagine that*...going for much longer before you can even manage to ejaculate...and *your unconscious knows how to increase the duration of pleasure*...and *you can become so absorbed in the pleasure you forget you are supposed to have ejaculated*...and I want you to start slow and not get too absorbed into the experience of the pleasure because I don't want you going too long yet before ejaculating...and you can *focus all of your attention on all the ways you can please the other person and get pleasure from pleasing the other person* and really try hard not to get too drawn into the experience of pleasure and forgetting to ejaculate...and your unconscious can now take as long as it needs over the next few minutes to honestly and fully integrate all this information at an instinctive level while I quieten down in the background...(pause for about 3 minutes)...'

Overcoming insecurity

'That's right...and as you continue to relax you can begin to create a deep and meaningful change...and that change can develop from the inside out...and with each breath you take that change can continue to grow...and you can wonder what it is like to live in security to live with a sense of feeling secure in yourself...and living in security where previously there was insecurity can be an experience...and your unconscious can begin to make the necessary changes on an instinctive level to enhance that sense of feeling secure in who you are as a person...and you can learn how to feel secure in relationships and what it is like in secure relationships...and secure people in secure relationships can develop securely...and a deep part of you can develop all the necessary changes that will ensure *you are secure in yourself and in your relationships*...and that you *remain calm and clear thinking*...and you can explore what life will be like when...*you feel secure*...in yourself and in relationships and with those around you...and you can explore what you will feel like in situations that in the past would have made that old you feel insecure...and you can notice what you say to yourself...and notice how you behave and what others will see and hear that would let them notice...*change has happened*...and what other differences does this change make to your life as you continue to...*grow up*...and...*develop*...and...*discover this new you*...and as you take some time to honestly and fully explore how feeling secure and being a secure person changes your life for the better...the instinctive part of you...that part of you that learn to respond in certain ways all by itself...that part of you can begin to take as long as it needs over the next few minutes to update fully and honestly all of your future response and make all the necessary changes to...make those future changes reality...(pause for about 3 minutes)...'

Heal ulcers

'That's right...and you can begin to learn how to soften on the inside...and relax...and as a part of you learns how to...manage stress in a new way...you can wonder what changes you will notice...and a part of you can review your diet and what changes are needed...and in studies when people have had the option of eating anything they want and they get that insight that certain foods or drinks lead to unpleasant effect they instinctively stop consuming those foods or consume a correct moderation of those foods...and when you experience stressful situations I wonder whether you will naturally discover yourself breathing in a new way or whether you will discover the muscles in your body *instinctively relax comfortably*...and your unconscious mind can send soothing and healing to where that is required...and that soothing and healing can take the time it needs to wrap up and protect that area and treat that area with your own internal natural healing processes...and you can begin to learn on an instinctive level how to *prevent the occurrence of ulcers in the future*...and you have a life time of experience of many example of times when you haven't had ulcers and your unconscious can learn what is different about those times...and how it can apply all that knowledge to the present and can give you a present here and now...and your unconscious can now take as long as it needs over the next few minutes to learn and integrate into the instinctive part of you all the necessary changes to help you...change quickly...and help you to know how to respond in the future...(pause for about 3 minutes)...'

Enjoy everyday sensory experience

'That's right...and as you wake up in the morning you can wonder how much you actually experience...imagine what the world would be like if viewed for the first time...and what those first few moments in the morning would be like as you open your eyes and perhaps notice how bright the light is...and notice the subtle shades and colours around you...and imagine how enhanced life would be if your senses were turned up...and you know it is possible to...*heighten your senses*...as if on a scale of 0-10 normal waking state of awareness is at a 5...and you can *turn your senses all the way up to 10* in pleasurable and everyday experiences...fully and honestly *enhancing your own personal daily pleasure*...and your sense of touch can increase in sensitivity and your sense of hearing and your sense of vision...and I don't know what differences you will notice...will sounds be clearer...will you hear a wider range of sounds...will you be able to notice individual notes...will you be able to recognise different smells...and appreciate finer detail in smells and tastes and colours...and I wonder what other changes you will notice as you begin to...*enjoy everyday sensory experience*...more fully...and really...notice the difference...as if *every day you wake up intensely happy to be alive*...and you can truly and honestly explore the power of your mind in ways you could only dream...and all the changes can happen with intense pleasure...and you know some people take drugs to enhance their senses...and all these drugs do is activate what is already there...and you can activate what is already there using the power of your unconscious mind...and really *explore new possibilities and change*...and you can be curious how much enjoyment you will get out of this...and I remember a boy that was hypnotised to increase his sense of touch and he ran his fingers over a smooth sheet of glass and thought it felt like sandpaper...and he then had his sense of smell increased and was asked what was in the hypnotists hand and he could smell the mint in the hypnotists hand from the back of the room...and there is even more...you can achieve now...and in a moment I'm going to quieten down and when I do your unconscious can make all the necessary changes to...make this happen...and it can happen fully and honestly in accordance with who you are as a person and in a way that enhances and benefits your life and the lives of those around you...and you can take as long as you need over the next few minutes to fully and honestly make all those changes on an instinctive level...now...(pause for about 3 minutes)...'

Overcoming procrastination

'That's right...and they say the hardest part of...*doing things*...is actually to just...*get up and just do them*...and we all have something that motivates us...something that makes us think I'm just going to...*get on and do it*...and as you think now about what you feel is a pleasure to do...and get a sense of where that image is in your mind...and get a sense of whether you are viewing the image as if you are there...seeing what you see...hearing what you hear...and feeling what you feel...or whether you are looking at that image and can see yourself in the image...and you can get a sense of how bright or dim the image is...how much contrast and colour is in the image...whether the sound is clear and as you hold that motivational image in your mind you can begin to notice a small dot in that image and rapidly as you notice the image disappear that small image of what you find you procrastinate about will rapidly enlarge and come forwards to fill this image...and all that will be changing is the image...in the same way that when you change the channels on a TV all that changes is the image...the setting for the sound...the colour...the brightness and all the other setting remain the same...and you want all those motivational setting to stay the same as you watch that image shrink down and disappear with the new image appearing almost instantly...and you can do that...now...flashing up that image that started as a dot while the main image shrinks...that's it...and now really take some time to look at this image and notice how it makes you feel different...more *motivated to take action*...and I wonder what else makes you...more *motivated to take action*...and you know how to...*start new things*...and find yourself...*compelled to carry them out*...and you have a life time of experience of...*achieving new things*...and I wonder how you did...and your unconscious can draw upon all of the resources you have used throughout your life to achieve things...and your unconscious mind can agree to stop that old procrastination response...and to signal that it has agreed it will open your eyes at the end of this session...and it has the whole session to think about whether it agrees before making that decision...and it will only make that decision if it agrees to *honestly and fully stop that old behaviour*...and it can find ways for you to step back and put things into stages and to prevent any feelings of feeling overwhelmed...and it can increase the need to *start new things and fulfil tasks*...and your unconscious knows what sort of a person you would love to be and how *you will become that person*...and you can take some time to explore in your mind what will be different in the future...who will notice you are getting on with things and achieving things...what differences will it make to your life...what positive differences will it make to the lives of others...and you can...*achieve things for yourself*...and while your unconscious take all the time it needs over the next few minutes to fully and honestly integrate all the necessary changes into the instinctive part of your mind I will quieten down in the background...(pause for about 3 minutes)...'

Enhancing concentration

'That's it...and you can discover ways to...*enhance your concentration*...and as you begin to get into a situation you want to...*concentrate fully*...in you can find yourself instinctively shutting out distractions and gradually focusing in more and more on what you are doing...and it is interesting how the more you focus on paying close attention to every little thing you do the more you shut other things out and...*enhance concentration*...and there is no need to try to...*enhance concentration at those times you need it*...because for concentration to occur fully...*it just develops by itself*...and whilst...*that concentration develops by itself*...you can just focus your attention on finer and finer detail about what you are doing...and if you are reading you can focus on what that voice in your mind sounds like...and how the eyes move...and what they feel like as they move...and what speed that reading is occurring at and if you are reading a book...what do the pages feel like...what do they feel like and sound like to turn...and you can focus down onto finer detail...and as you do...*your concentration will increase*...and I don't know whether you will decide to do that consciously or whether it will happen unconsciously...and in appropriate situations you can prepare in advance to concentrate...you can get the mind-set first and set up the situation and make sure the situation has as few distractions as possible and take a few moments to really focus internally perhaps on breathing or the feel of the air as it flows through the nose...then when you have prepared...*you can concentrate fully*...and I'd like to have you take a few moments just to think about those times you have entered that state of concentration...even those times it was for just a few seconds...and there are thousands of times in your life this has happened *automatically* without you being aware of it...and you can begin to gather up all of those times and learn from them how...*your unconscious knows how to concentrate fully*...and it can be instinctive...and automatic...and you can begin to get a sense of how you will know...you are concentrating fully in the appropriate situations in the future...if I was a fly on the wall watching you concentrating how would I know...what would I see...what behaviours...what body language...what will I hear...how does...concentrating in those times in the future...make thing different...what are the positives of that...how does it impact on your life and the lives of those around you...and you can now take as long as you need over the next few minutes to fully and honestly make all the necessary changes on an instinctive level to ensure you can easily and effortlessly concentrate in the necessary future situations...(pause for about 3 minutes)...'

Focus meditation

'That's right...and you can now begin to prepare to experience a focused state of meditation and in a moment a single short word or letter or sound will come to mind and when it does you can begin to focus on it...and you can begin to focus all of your attention on that sound...and once that sound is in your mind you can begin to say that sound to yourself and you can stretch that sound out to fill your out breath and allow your mind to be empty on each in breath...and as you make that sound to yourself in your mind you can...*just observe...observe without judgement*...and just observe the full range of that sound...what colours come to mind...how the sound reverberates...how the sound is associated with certain feelings...and really just...*observe what comes up in your mind* yet focus on ignoring all distractions and associations and focus on repeating that sound with each out breath...and continue that...and if your mind wanders bring it back on track and...*focus on that sound*...and with each breath out you can become...deeper and deeper absorbed... (pause for about 5-10 minutes)...'

Loving kindness meditation

'That's it...and just begin to get a sense of the heart beating...and begin to notice that it isn't just blood being pumped round the body...notice how there is a healing loving light flowing through the body and mind...and notice how this energy is pulsating with each beat of the heart and beginning to radiate out from the self...and as you observe that energy flowing out notice how that loving kindness begins to interact with the energy of others...it begins to feed in pleasure and healing and a connection with those that it touches...and get a sense of that loving kindness being transmitted into the hearts and souls of every living being...allowing you to see and connect with the inner beauty of others...and in a moment I will go silent and when I do the transmission of healing loving kindness can intensify...and you can begin to absorb the loving kindness given off by others...and you know what it feels like to be truly and unconditionally loved...and as you begin to recall that now while that heart continues to beat you can intensify that feeling of loving kindness...and you know what it feels like to love truly and deeply...and as these thoughts of all the times in your life you have felt a deep sense of love and all the times you have felt deeply loved you can take that feeling...feed it into your heart and allow it to be pumped out into the world around you...and this loving kindness that permeates you and spreads out from you can continue indefinitely...and you can see inner beauty...and there is a story about a Buddhist monk...and he went to a temple and took his new expensive shoes off at the door...after he had finished praying he returned to the door only to notice his shoes were missing...the monk calmly left the temple and went home...when he arrived he was asked where his shoes were...and the monk replied that they were taken...he was asked didn't that make you angry...no...replied the monk...I can only apologise to the thief that I have created...had I not worn such expensive shoes I never would have tempted a less fortunate person to feel the need to steal...and you can discover how every unkind act leads you to feel a sense of compassion and understanding and loving kindness to others...*and silence can intensify that*...now... (pause for about 5 minutes)...'

Overcoming nail biting

'That's it...and as you continue to relax you can begin to learn new ways to respond...and up until now you had felt it was necessary to eat those nails...and you hadn't felt it was necessary to...*enjoy a new experience*...and a part of you can drift back into the past to a time before that problem began...and as it drifts slowly back forwards in time it can settle on all relevant memories and begin to...*remove all emotion from those memories*...and I don't know whether the unconscious mind will do this with the conscious mind knowing or without...and that part of you can then continue forwards until it reaches the present...which is given in the here and now...and in the future...you can begin to experience memories where in the past you would have ended up biting those nails...and you can notice how *this time things go differently*...and I wonder what will be different and new...and you can notice how calm you can be in those future situations that in the past would have led to that old nail biting behaviour...and you can mentally rehearse dozens of future situations that in the past would have led to that old nail biting behaviour...and notice what is different...and notice how you know...*you are responding differently*...and what you can see and hear and what it feels like to be calm responding differently...and what do others notice...how do they know *things are different*...and you can really learn on an instinctive level from this experience...and find that in the same way that a dropped plate that breaks doesn't suddenly jump back and fix itself neither does that old behaviour when...it becomes this new response...and while I quieten down in the background your unconscious can programme all the necessary changes into the instinctive part of your mind...and you can take as long as you need over the next few minutes...(pause for about 3 minutes)...'

Change that problem now (generic problem resolution script)

'That's it...and as you think about what life will be like in the future...now...when that old problem has gone you can explore what that is like...and as you explore what that is like I wonder what will be different...and you can...notice those differences and notice what life is like without that old problem...and you can get a sense of what you do differently...what will someone else see if they are watching you...how do you feel...what does someone else hear...and how do those changes affect other people in your life...and you can get a sense of being a long way off in the future looking back and reminiscing on how you achieved success and overcame that old problem...and whether you had any brief relapses or just remained problem free...and while I go silent you can take as long as you need to really fully and honestly explore how that changed occurred...what led to you overcoming that old problem...how long it took you to fully overcome that problem and whether there were any difficulties that you had to overcome...on the road to success...and all those changes and all that is relevant can become a part of the instinctive part of you by the time I talk again and you can also drift back from that future position at the same time as...finishing all that is necessary to overcome that problem...(pause for about 3 minutes)...'

Warts removal

'That's it...and you can get a sense of a vast land...and on that land are a number of invading forts...and you as the commander of the army have been given the task of destroying those invaders...and you can *learn from this on a deep unconscious level* and put into practice what you learn...and there was a case in the past where a hypnotherapist thought a boy had an extreme case of warts and the hypnotherapist removed the warts before discovering that it was actually a genetic disease and should have been impossible to cure...yet...*the cure happened*...and there was a case of someone being thought to have hysterical blindness...and the therapist used hypnosis to treat this...after a number of sessions the man was beginning to see again...and when he went to a neurologist it was discovered that he actually had brain damage and so should have been impossible to cure yet he was now seeing...and hypnosis has been used for centuries to cure warts...and sometimes the hypnotist has just suggested...*those warts will vanish*...the skin there will grow back clean and clear and new...*and the warts have vanished*...other hypnotists have been a bit more vague and suggested...*those warts will vanish*...and I don't know whether it will happen this week or next week or within a month...and...*the warts vanish*...and others will tell stories and the clients unconscious mind picks up on the patterns and understands the stories on a deep level and makes those changes to...*get rid of the warts*...and I don't know which option you will respond best to...and it can *happen automatically*...and you know the commander can send in an army to attack the invaders and over power them...or the commander can stop the food supply getting to the invaders so they gradually run out of fuel and die...and I don't know what option you will choose...and how will things be different once they are gone...and I don't know whether it will be as if...*they just fall off*...or whether it will be as if...*they fade away*...and...*change happens*...all the time...it is just a matter of how and when...and I wonder how you will behave differently...and if there are any underlying causes behind those old warts being there then a part of you can drift back and tackle those underlying issues...if there are any memories associated with the creation of those warts...*your unconscious*...can remove all emotion from those memories whilst maintaining any important learning...and that part of you can go all the way back outside of conscious awareness to the beginning...to when the warts first appeared and what specifically caused them and how you can...now move forwards...and you can take as long as you need over the next few minutes to make all the necessary changes on an unconscious instinctive level...and be curious how quickly change will begin and what you will first notice...(pause for about 3 minutes)...'

Help with meeting basic emotional needs and using their innate skills correctly

To give and receive attention

'And while you continue to relax I wonder what you are thinking...and as you think those thoughts a part of you can begin to...*create change*...and *your unconscious*...can begin to deeply consider new possibilities about how...you will find ways to honestly compliment as many people as you can each day...and I don't know whether it will be three people or five people or seven people or more or maybe even four...and each day you can find something you can do that will make someone else feel good without doing it to get anything out of it yourself...and it can be something as simple as smiling more or holding a door open or asking 'can I help you with that?' and helping...and you can make all the necessary changes and decisions to allow yourself to take some time to do this...(pause for about a minute)...'

Mind body connection

'And as you relax you can continue to think those thoughts and a part of you can begin to...*create change*...and *your unconscious*...can begin to deeply consider new possibilities about how...you will pay attention to the mind and body...and the mind and body are interconnected and influence each other...and when you inhibit sleep your intelligence and ability to carry out mental tasks reduces...so what is happening to your body affects the mind...and when you worry excessively and get stressed you struggle to sleep...and so your mind affects your body...and when people take drugs or alcohol or smoke the chemicals affect the body and these changes change the mind...and when people suffer depression or anxiety these create chemicals that change the body...and when you eat healthy the body stabilises...blood sugar levels stabilise...other internal chemicals and processes stabilise and work more efficiently...leading to greater alertness and clearer thinking...and when we laugh and find things funny and think about things that make us happy that releases the body's natural opiates and feel good chemicals into the blood and creates many other positive physiological changes...*making the body healthier*...more likely to reduce sensations of discomfort...*improve sleep*...improve health and wellbeing and improve the ability to handle situations...and I don't know what changes you will make to make sure the mind and body develop a healthy interaction and a healthy influence on each other...and you can wonder also how this will happen or just discover it happening all by itself...and a part of you knows what changes you as an individual will make to...*create change now*...(pause for about a minute)...'

Purpose and goals

'And you can discover something previously unknown about a future change and whilst you are here in this state of mind you can be a key catalyst of the new change...and you can begin to plan things in the future...and I don't know whether you will plan what you will do each day or what you will achieve each week or just what you will achieve in a month or year...and you can...now...*discover what you are good at*...and you may never have previously even realised...and a part of you can *discover that now*...and I don't know whether that part of you will allow the conscious part of you to share in its discovery or if it will wait a while before sharing that discovery...and your unconscious can begin to get a sense for how...*you can start to write lists about what needs to be done each day or what you would like to achieve* each day...week...month...or year...and as...*you begin to achieve*...you can mark off all the successes and see just how much of that...*you will achieve*...and I wonder whether...*you will achieve*...50% of what you set out to achieve...or 75%...or 100%...or another amount entirely and I wonder whether...*you will begin to achieve*...and then...*discover you change your mind*...and...*set new goals and targets*...or whether you will just stick to what...*you set out to achieve*...and I don't know what types of things...*you will set to achieve*...whether it will be everyday tasks like...tidying up...planning the weekly shopping...planning what to eat at meal times...or whether it will be things like...achieving at work...like setting what needs to be done daily and ticking off each item as...*you achieve* it...or maybe setting a dream that you will achieve and how...*you will achieve* that...and when by...and setting a time scale for each achievement...and ticking off each stage as...*you achieve* it...and you can take some time to integrate this new way of thinking into the instinctive part of your mind...(pause for about a minute)...'

Belonging to the wider community

'As you continue to drift and dream you can wonder about how you can enjoy getting something new out of life...and the part you play in the world around you...and for a tree to grow it needs good soil and water and nutrients and sunlight and other plants around it and animals...and without these the tree doesn't grow up big strong and healthy...and it doesn't just need any plants or any animals or any amount of sunlight or any nutrients or any type of soil...it needs to ensure that the soil is just right for the type of tree it will become...it needs to be in a location so that it will have just the right amount of water and sunlight...and it needs plants around it that work with it so that they get the right amount of support and shelter whilst offering something back to the tree in return...they will supply the tree with appropriate nutrients and protection and will work together...and the animals need to also help the tree...they need to be drawn to the tree for what the tree has to offer and in return they take on board seeds and pollen and freely and willingly carryout tasks for the tree...and together they support each other...and your unconscious can understand on a deep and meaningful level what I am saying...as you take some time now to...integrate all that...(pause for about a minute)...'

Stimulation and creativity

'That's it...and we all need to...*feel a sense of stimulation*...and we all need to...*be creative*...and it doesn't mean we all need to become artists or musicians or to fill our lives will exciting experiences...it is more fundamental than that...we all need to find ways to...*feel a part of the world*...around us...and to feel we all...*make a contribution to the world*...around us...and creating stimulation can be as simple as taking time to read every day or to learn something new every day or to fill your day with a number of activities...if you sit in a room all day and do nothing you can get bored...if you sit in a room for a few hours a day and spend the rest of the day going out for a walk...reading a few chapters of a book...listening to some music...watching a programme on TV...meeting up with a friend...playing a game for an hour or so...cooking a meal...phoning a friend or relative...planning a holiday...all of these are stimulation events...they are all small events that fill a day with stimulation...and there are many more you can think of...and it is important to *develop a variety of stimulation events* rather than just one or two and creating and thinking about stimulation events involves creative thinking and coming up with new ideas...and...*adapting to life*...and your unconscious can take as long as it needs over the next minute or so to develop these ideas creatively into your future and build on them...(pause for about 1 minute)...'

Understood and emotionally connected to others

'And you can learn something now about how you connect with others around you...and how you *help yourself to be understood*...and there are times you are more understood and times you are less understood and times *you feel close to people* and times you feel a little more distant...and your unconscious can integrate new learning and understanding about how you connect with people in a unique way...and sometimes it can seem like you or others are speaking a different language...and that you could do with a translator...and your unconscious can act as a translator and you may not even realise...and we communicate on many levels and to be understood we have to pick up on many levels of communication...and as you listen to people you can really focus and learn and develop with what they are communicating to know how to put what you want to communicate into a way they will understand...and you can understand through communication and discovery and learning...and when you are around someone that spends time breathing like you and spends time using similar words to you and similar categories of words to you and spends time smiling yet much of this is so subtle you don't consciously notice you can find yourself *feeling deeply emotionally connected* and feeling all of your senses and emotions lighting up with pleasure...and you can *form appropriate emotional connections with those around you*...and you can learn to recognise the signs that allow you to know when to approach and when to back off...and when you approach a wild horse you know you are no threat and you know you just want to get close to them but if you approach too fast or approach looking aggressive or threatening the horse runs away...if you take time and approach at the right speed paying attention to the horse so that if the horse begins to show any signs of being uncomfortable you slow down...if they appear more comfortable you can speed up...until eventually you reach the horse...and you know if a horse approaches you for food or for petting you should feed or pet that horse and even horses can make the mistake of pushing their luck even though they have been fed or petted...and you can learn more from this than you realise on a conscious level...(pause for about one minute)...'

A sense of control and independence

'That's right...and it is natural to want to have control...and sometimes people can feel too in control of what they have very little control and other times people can feel too out of control of what they actually have more control over...and you can begin to learn about how much control you have in different situations...and some people crave control and suffer when they can't control those things you have no control over and others wish they had more control and feel they don't have control because they are trying to control those part of the experience that they have no control over...and if someone jumps out of a plane with a parachute on they have no control over the falling...they will continue falling...and if they want to try to control that decision they will get stressed and fail...they do have control of how they fall...where they fall...how fast or slow they fall...and when they slow the fall down...and they can pull the cord and release the parachute making the fall comfortable and controlled...and survivors in prisoner of war camps had no control over whether they were tortured or not...and the survivors all took control which helped them survive...they took control of what they could control...they took control of how long they went before screaming or what they spent their time thinking about...and they often thought about getting home to see loved ones...and too much control leads to problems forming...like obsessive problems or superstitions where people try to control things they have no control over and then the problem gets out of control and so they try to control things more...and by letting go and thinking...do I have real control here...and only controlling what is in your control...leads to an *improved future*...of change and discovery...and you can control how you act and behave and you can't control other people or inevitable events...and you can focus you controlling efforts on those areas you can control and control them in a way that positively benefits you and those around you in a way that enriches and enhances lives...(pause for about a minute)...'

Tolerating uncertainty

'And as those eyes move in that way you can begin to explore ways you can become comfortable with a certain amount of uncertainty where it is appropriate and explore those times there has been uncertainty and pleasure and you can learn on an unconscious level how to increase comfort in situations where you don't know the outcome or how it will go or where you are only in control of yourself and how you react in that situation...and you don't want to react in uncertain way that makes it appear you have lost control and you can react in a way that displays true honest unconscious control...and you can programme that into your neurology now...(pause for about 2 minutes)...'

Helping the client challenge their emotional thinking

'And as you continue to listen to me you can begin to learn how you can...*step back in your mind*...and how you can...learn to *step back in your mind* in future situations...and how this...*stepping back in your mind*...can allow you to get a new perspective on a problem...and in situations in the future you can...*step back in your mind*...and see the wood for the trees...and you know it is easy for a football fan watch football on a TV and shout and the TV and know the player should have passed the ball...yet the player who is the expert didn't notice the opportunity because they were in the game and too busy focusing on not losing the ball...and you can learn from this how you can...*change your mind*...and develop in new ways...(pause for about 2 minutes)...'

Transferring natural resources, skills and strengths

'That's right...and you have all the resources and skills you need at an unconscious level based on all of your life experiences and as you relax here and...now...you can begin to...*learn unconsciously*...how to link skills...resources...and abilities from one area of your life to other areas of your life...and you don't have to know how...*you do this*...it can happen automatically...as you take time to drift deeper and deeper into this state...(pause for about 2 minutes)...'

Managing attention

'*Deeper and deeper*...that's it...and you know some things make us become absorbed while other things we just find we can't concentrate on...and as you listen to me talking you have been learning how to...*manage attention*...on an unconscious level...and you know if there was an emergency your attention would be fully on that emergency...and you know on an instinctive level how to...*manage your attention*...and can really master...*managing your attention*...and you can discover that changes to how...*you manage your attention*...occur automatically over time without any effort on your part...and you can take time now to let that integration take place fully and honestly...(pause for about 2 minutes)...'

Using creative thinking and thinking of alternative viewpoints

'And you know there is often more than one meaning to an event or situation and your unconscious can instinctively begin to offer you different meanings to situations that could in the past have caused distress...I remember the story about the driver that was driving along the motorway when someone cut them up...they were fuming but then they started laughing....and another person was driving down a motorway and they got cut up and were fuming...and then they behaved differently...and on a quiet country road a Porsche and a Ferrari crashed into each other causing lots of damage to the vehicles...yet the owners started laughing...and why did people respond in such odd ways...well the first person to be cut up saw the driver of the car in front wave and recognised her as a good friend...the second driver that was cut up noticed what looked like a distressed pregnant woman in the back seat of the car and was now more concerned that the person gets to the hospital quickly and the two expensive cars that crashed turned out to be a brother and sister and the brother had bought the sister the Ferrari as a present and they saw the funny side of the incident as no-one was hurt and it was at a slow speed...and you can now take time to learn on an unconscious level how to hold multiple view points and understandings and different meanings of situations and how different meanings can lead to different feelings and different reactions and decisions...and you can take some time now to work out how this applies best to you and how this can help you to improve in the future...(pause for about 2 minutes)...'

Reorienting Back to the Room

Exiting trance now

Generally I would only tell the client when to exit the trance if we were short of time in the session. Normally my preferred option is to allow the client to exit trance when they are ready rather than because I told them to. This also allows for the therapy and the client deciding to open their eyes to be linked together so that you can suggest something like 'and when you know instinctively that all the work is complete you can drift out of trance and open your eyes' this then links the eyes opening with acceptance of all the work being done.

'That's right...and you can now just take a few moments to allow yourself to get a sense of drifting back into that chair...and I don't know whether you will take 4, 5, 6 or 8 breaths to fully re-orient back to the chair...before you then start to work your way back along the route you took into trance...and as you do your unconscious mind can ensure that all the new re-programming has been installed throughout the mind and body in every cell and neuron...your unconscious mind can honestly and fully integrate all your new learning into that deep and instinctive part of your mind while you continue your journey back to the here and now...and I don't know how much of what is there will be left there as you continue back and how much you will bring back with you...an it's interesting how much you know you can be aware of while you are sat there in that seat...and you can now find yourself working all the way back away from that state to the here and now becoming more aware of sounds around you...of me of what you will be getting on with later and other random thoughts that start to cross the mind and in a moment I'm going to count to 3 and on the count of 3 you can open your eyes and be fully back in the room...one...two...three...and open your eyes...hi now you came here to see how you are able to be helped and I asked you to close your eyes earlier...so do you have any plans for later...(have a brief chat then if you are going to set any tasks etc. you can do that before they go)...'

Allow the client to exit trance when they are ready

This is the preferred option as it gives the client control over the process and allows for the success of the trance work to be contingent on the client opening their eyes.

'That's right...and now as you gradually find your way back out of trance you can take all the time you need to ensure all the work is completed fully and honestly...and only drift back out of trance at the rate and speed that all of the re-programming updates...and when you have found your way back fully all necessary changes will have happened on an instinctive level allow you to open your eyes and come back to the room...(remain quiet while they take as long as they need to return to the room, once they are back in the room with their eyes open then say 'hi' and talk briefly about something irrelevant relating to before the session like how long it took them to get there then ask about if they any plan for the rest of the day, then set any tasks and have some general chat to make sure they are very much awake now)...'

Allow the client to drift off to sleep

This ending is included for those that want to make self-hypnosis tracks for clients and feel that the client should be able to listen to the track at night and just drift off asleep. Some clients prefer this option with self-hypnosis tracks rather than being woken from trance to then have to fall asleep; especially clients that already have difficulty sleeping.

'that's right...and now as you gradually...drift off to sleep...you can take all the time you need to ensure all the work is completed fully and honestly...and only drift out of trance and into a comfortable deep sleep at the rate and speed that all of the re-programming updates...and as your neurology updates you can find your way back fully out of trance and into that relaxing sleep...and when you drift into a dream...all necessary changes will have happened on an instinctive level...and you can awaken at the appropriate time feeling refreshed and revitalised...(the client can then drift to sleep)...'

Treatment Ideas

Gathering information

Hypnotherapy isn't all about the scripts. It's about good quality information gathering that can be useful to help move clients on. It could be that something you say or ask a client stimulates certain mental processes that lead to future change even if these questions weren't seen as part of the 'therapy'.

Many people I work with have often moved on considerable between talking with me and having a first session. Often by the second session things have moved on tremendously with the client having insight into their situation and understanding how to move forward resiliently.

Sometimes you need to find out what purpose the current problem is serving and work with the client to find an alternative solution (not all problems are currently serving a purpose but many are so it is worth checking out). From this discussion it may turn out that the client needs to learn something like how to relax or ways to manage stress before they will be in a position to move on.

If this is the case you can explain this to the client and how this is part of the treatment plan. If you do this you can set a goal of the session for example to be to learn to relax and practice a specific technique between now and the next session. Then you can teach them a technique or skill and just give them a brief hypnotic experience.

You can then ask them in the next session what changes they have noticed and explore any changes they report. By getting a clear understanding of the client's problem and the solution they want you can formulate your own ideas and suggestions that are specific to that individual.

When gathering information ask 'who, what, where, when, how' questions. You want to know as much information as possible about the structure of the problem and the aspects of the problem like does it always happen around the same people or in the same situations. You also want to know about the solutions so when doesn't the problem happen, where doesn't the problem happen, who doesn't the problem happen with.

Look for exceptions as these are often where you will find the solutions. Normalise the problem for the client so that they don't feel they are the only person that has that problem. Scale the problem at the beginning (E.g. from 0-10) and then at the end. This will allow you and the client to see progress. Find out what the number they give means in sensory language. 'What does a three look like?' Then ask what stops it being one point lower and what would make it one point higher. And where would they ideally like it to get to and what would all this look, sound, feel like, who else would notice etc.

Ask the client for resources. Some of this would come from asking for exceptions but they may also have skills in other areas of their life like being able to be calm at work but not at home. So you want that resource and can then explore how they can transfer it from one context to another and explore what is different that makes them able to use it in one context and not the other. Resources could also be family and friends or hobbies. Things that can be used to tackle the presenting issue.

Ask hypothetical questions like 'If you woke up tomorrow and you no longer had the problem but no-one told you what would be different. What would you see, hear, feel. What would others see and hear or notice?' From this clients can begin to create solutions and begin to develop a mindset of noticing improvements and can also bring out more useful information and resources.

Task setting

Tasks can be a good way of helping the client to take what they are getting in the session out into the real world. Many clients do well in sessions but don't do anything between sessions because they don't realise that they can.

Tasks can take on a number of forms. You can set metaphorical tasks. These are tasks that have no fixed meaning and are vague so the client finds the meaning themselves. As long as they are issued with sincerity they can be very effective. For example you could set a task of learning about cacti as a way of teaching a client that they don't need to smoke (cacti can cope perfectly well in harsh environments with limited water which could be a metaphor for someone coping well in a stressful life without resorting to smoking). It could be having the client find a specific shape and sized stone on a beach. Carry it around with them for a week and wonder what it means and before throwing it away at the end of the week (could be a metaphor for throwing away an old habit).

Metaphorical tasks can also work well if the metaphor was used in the trance work. So for example if you used a metaphor like comparing a congested city to a clean and clear countryside you could have a task where they spend a day in a city somewhere noisy and busy and just observe and then a day in the country; making the city experience unpleasant and the country experience pleasant.

You can set tasks that alter the problem pattern in some way. It could be altering the time the problem takes place. So if they always smoke when they answer the phone you can set the task of having them smoke once the phone call is finished. Any change to an established pattern begins to disrupt it and the longer the disruption goes on for the more impact it has on the pattern. In many cases if you can convince a client to alter their problem pattern and continue with this alteration the problem often disappears.

You could alter the location the problem takes place. So if someone smokes in a specific location you could have them smoke in a different location. You can alter the length of time the problem behaviour is carried out for. So if someone takes three minutes to smoke a cigarette you could get them to agree to take 30 seconds or one minute or five minutes or ten minutes. This alteration is likely to become a chore and begin to add in extra feelings of not wanting to do it. If you have got the client to agree to continue whenever they smoke then their only option is to stop smoking.

You could add in extra stages or change the order of the stages that are already there. Like having to run every time someone wants a cigarette; or having to count to twenty between each drag on the cigarette. These extra stages again create a chore and link a feeling of not wanting to do the behaviour. Even in situation where it isn't a chore it still disrupts the pattern meaning the pattern will never be the same again.

Getting emotional needs met

We all have emotional needs that need to be met to lead an emotionally balanced life. These needs include giving and receiving attention, having a sense of control in our lives, having purpose and meaning in our lives, having stimulation and creativity; and being aware of the mind body connection.

If we don't have needs being met appropriately we may seek out alternative sources to get them met. So for a problem like smoking it could be smoking as a way of 'breaking the ice' when socialising. Or smoking as a way of making sure we take a break (to meet our mind body connection need). Or smoking to have a sense of control in our lives as everything else seems out of our control yet we can decide to smoke and when to smoke; or smoking to give us stimulation. So just one problem can have many different ways it needs to be treated.

If you can learn what need is being met by the problem that the client is doing you can then address that need directly and as that need gets met appropriately in another way they automatically become less likely to continue with their problem and are more likely to find it easier to move on and keep their changes.

You can also set tasks that will put the client in situations where they will get their needs met appropriately and learn new ways to meet those needs. Or encourage clients to take up hobbies or interests that help them to meet their needs.

By listening out for needs that aren't being met or are being met inappropriately you can dramatically speed up the time it takes to treat clients because you can have a greater impact on their life and on how they will continue to cope with life after therapy.

Made in the USA
San Bernardino, CA
18 April 2017